PRACTICE PROBLEMS IN NUMBER SYSTEMS, LOGIC, AND BOOLEAN ALGEBRA

SECOND EDITION

by Ed Bukstein

HOWARD W. SAMS & COMPANY
A Division of Macmillan, Inc.
11711 North College, Suite 141, Carmel, IN 46032 USA

© 1967, 1973, and 1977 by Howard W. Sams
& Co.
A Division of Macmillan, Inc.

SECOND EDITION
FOURTEENTH PRINTING—1990

All rights reserved. No part of this book shall be
reproduced, stored in a retrieval system, or transmitted
by any means, electronic, mechanical, photocopying,
recording, or otherwise, without written permission
from the publisher. No patent liability is assumed
with respect to the use of the information contained
herein. While every precaution has been taken in the
preparation of this book, the publisher assumes no
responsibility for errors or omissions. Neither is any
liability assumed for damages resulting from the use
of the information contained herein.

International Standard Book Number: 0-672-21451-2
Library of Congress Catalog Card Number: 77-72632

Printed in the United States of America.

PREFACE

The areas of number systems, logic, and Boolean algebra have gained vastly increased importance because of the great strides being made in digital computer technology. The increasing utilization of computers in business, commerce, industry, and science has created an increasing demand for people knowledgeable in these basic subjects.

This workbook is a 64-lesson introduction to digital computer mathematics. The first lessons cover the various number systems, e.g., binary, trinary, octal, and decimal, converting numbers from one system to another, and some common codes. The binary and octal arithmetics are then developed, laying a basis for the introduction of Boolean algebra. The latter, with its relations of AND, OR, and NOT, is elaborated by truth tables, Venn diagrams, Karnaugh maps, and Veitch diagrams. The implementation of Boolean algebra in electronic gating and inverting circuits also is covered.

The problems in this workbook are designed for use either in the classroom or by the individual student. No special mathematical background is required. In the classroom, the workbook may be used in its entirety, or the teacher may select those topics that best suit his time schedule and desired depth of coverage. Answers to the assignments are given at the end of the book.

ED BUKSTEIN

CONTENTS

1. Number Systems 7
2. Counting 9
3. Octal-to-Decimal Conversion by Positional Values 11
4. Decimal-to-Octal Conversion by Remainder Method 13
5. Conversion of Fractional Decimal Numbers to Octal Equivalents 15
6. Binary-to-Decimal Conversion by Positional Values 17
7. Conversion of Fractional Binary Notation to Decimal System 19
8. Decimal-to-Binary Conversion by Remainder Method 21
9. Conversion of Fractional Decimal Notation to Binary System 23
10. Binary-to-Decimal Conversion by Double-Dabble Method 25
11. Octal-Binary Conversion by Substitution 27
12. Binary-Coded Decimal Notation 29
13. Excess-Three Code 31
14. Two-Out-of-Five Code 33
15. Miscellaneous Codes 35
16. Alphanumeric Codes 37
17. Addition of Binary Numbers 39
18. Binary Subtraction by Direct Method 41
19. Complements 43
20. Decimal Subtraction by Complement Method .. 45
21. Binary Subtraction by Complement Method ... 47
22. Binary Multiplication and Division 49
23. Octal Addition and Subtraction 51
24. Octal Multiplication and Division 53
25. Hexadecimal Number System 55
26. AND/OR Relations 57
27. AND/OR Logic 59
28. Logic Symbols 61
29. Boolean Postulates 63
30. The NOT Concept 65
31. Theorems 67
32. Removing Common Factors 69
33. The Truth Table 71
34. Construction of Boolean Truth Tables 73
35. Use of Truth Table 75
36. Converting Boolean Equations to Block Diagrams 77
37. Converting Block Diagrams to Boolean Equations 79
38. Converting Block Diagrams to Truth Tables .. 81
39. Converting Truth Tables to Block Diagrams .. 83
40. Converting Boolean Equations to Truth Tables 85
41. Converting Truth Tables to Boolean Equations 87
42. The Venn Diagram 89
43. Converting Venn Diagrams to Boolean Expressions 91
44. Converting Venn Diagrams to Block Diagrams 93
45. Converting Block Diagrams to Venn Diagrams 95
46. Repetition of Boolean Expressions 97
47. Some Useful Relations 99
48. Simplification of Logic Diagrams by Use of Basic Theorems 101
49. $\overline{AB} \neq \overline{A}\overline{B}$ 103
50. De Morgan's Theorem 105
51. Application of De Morgan's Theorem 107
52. True-Output and False-Output Forms 109
53. Inverting Boolean Equations 111
54. Minterm and Maxterm Equations 113
55. Karnaugh Maps 115
56. Construction of Karnaugh Maps 117
57. Adjacencies in Karnaugh Maps 119
58. Readout of Karnaugh Maps 121
59. Simplification of Boolean Equations by Use of Karnaugh Maps 123
60. Simplification of Logic Diagrams by Use of Karnaugh Maps 125
61. Inverting Boolean Equations by Use of Karnaugh Maps 127
62. The Veitch Diagram 129
63. Additional Notes on Logic Circuits 131
64. Additional Notes on Logic Symbols 133

Name_____Date_____Class_____Grade_____

1. NUMBER SYSTEMS

The *base* or *radix* of a number system specifies the number of symbols available in that system. A number system based on the letters of the alphabet, for example, would have a radix of 26. The decimal system has a radix of ten because it employs ten symbols: 0 through 9. Eight symbols (0, 1, 2, 3, 4, 5, 6, and 7) are employed in the *octal* system, and the *binary* system has only two symbols (0 and 1).

In the decimal system, digits to the left of the decimal point represent ones, tens, hundreds, thousands, etc. These *positional values* or *weights* can be expressed as powers of ten: 10^0, 10^1, 10^2, 10^3, etc. Digits to the right of the decimal point represent tenths, hundredths, thousandths, etc. and therefore have positional values of 10^{-1}, 10^{-2}, 10^{-3}, etc. The decimal number 952.68 therefore represents:

$$9(10^2) + 5(10^1) + 2(10^0) + 6(10^{-1}) + 8(10^{-2})$$

The positional values of the octal system are powers of eight. Octal number 374.67 therefore represents:

$$3(8^2) + 7(8^1) + 4(8^0) + 6(8^{-1}) + 7(8^{-2})$$

Assignment: Answer the following questions. Place answers in the spaces provided.

_____ 1. In the octal system, what is the positional value of the fifth digit to the left of the octal point?

_____ 2. In the quinary system (radix = 5), what is the weight of the second digit to the right of the point?

_____ 3. In a number system based on the letters of the English alphabet, what is the weight of the third digit to the left of the point?

_____ 4. Which of the following represents the larger quantity: decimal number 12 or octal number 15?

_____ 5. A trinary number system employs three symbols (0, 1, and 2). Which represents the larger quantity: trinary number 222 or decimal number 25?

2. COUNTING

In a number system of any radix, counting is accomplished by advancing the least significant digit one unit at a time. In a number system based on the letters of the alphabet, for example, counting progresses as follows:

      aaa
      aab
      aac
      aad
      aae
      . . .

When the digit being advanced has reached the last available symbol, it next returns to the first symbol and the digit to its left advances one unit. Thus, in the alphabetic number system, *aaz* is followed by *aba* as shown below:

      aax
      aay
      aaz
      aba
      abb
      abc
      . . .

Counting proceeds in a similar manner in all orderly number systems. When it is not obvious from the context, subscripts are employed to designate the radix of the system in which a quantity is expressed. For example, 101_2 and 6173_8 are binary and octal numbers respectively.

Assignment: Answer the following questions. Place answers in the spaces provided.

    1. In the alphabetic counting system, what number follows azz?
    2. In the alphabetic system, what number follows azzz?
    3. In the octal system, what number follows 077_8?
    4. In the binary counting system, what number follows 01_2?
    5. What number follows 11111_2?

Name_____ Date_____ Class_____ Grade_____

3. OCTAL-TO-DECIMAL CONVERSION BY POSITIONAL VALUES

Numbers expressed in the octal system can be converted to the decimal system by multiplying the positional values (powers of 8) by the corresponding octal digits, and then adding as shown below.

Example: Convert 2417_8 to N_{10}.

$$2417_8 = 2(8^3) + 4(8^2) + 1(8^1) + 7(8^0)$$
$$= 2(512) + 4(64) + 1(8) + 7(1)$$
$$= 1024 + 256 + 8 + 7$$
$$= 1295$$

Therefore $2417_8 = 1295_{10}$.

Assignment: Using the method of positional values, convert the following octal numbers to their decimal equivalents. Place the answers in the spaces provided.

_____ 1. 10007_8 _____ 6. 26_8
_____ 2. 100_8 _____ 7. 64_8
_____ 3. 1234_8 _____ 8. 551_8
_____ 4. 777_8 _____ 9. 62_8
_____ 5. 4131_8 _____ 10. 45.5_8

Name_____ Date_____ Class_____ Grade_____

4. DECIMAL-TO-OCTAL CONVERSION BY REMAINDER METHOD

A number expressed in the decimal system can be converted to the octal system by successive divisions by 8. The remainder in each division is retained as a digit of the octal number, the first remainder being the least significant digit.

Example: Convert 483_{10} to N_8.

$$483/8 = 60 \text{ and remainder of } 3$$
$$60/8 = 7 \text{ and remainder of } 4$$
$$7/8 = 0 \text{ and remainder of } 7$$

Therefore $483_{10} = 743_8$.

Assignment: Using the remainder method, convert the following decimal system numbers to their octal equivalents. Place the answers in the spaces provided.

_____ 1. 27_{10} _____ 6. 512_{10}

_____ 2. 635_{10} _____ 7. 456_{10}

_____ 3. 2626_{10} _____ 8. 8888_{10}

_____ 4. 1007_{10} _____ 9. 206_{10}

_____ 5. 905_{10} _____ 10. 4095_{10}

5. CONVERSION OF FRACTIONAL DECIMAL NUMBERS TO OCTAL EQUIVALENTS

Fractional quantities expressed in the decimal system can be converted to octal by repeated multiplication by 8. The procedure is as follows: Multiply the fractional quantity by 8, then use the fractional part of the answer for the next multiplication by 8, etc. In the result of each multiplication, the digit immediately to the left of the decimal point is retained as one of the digits of the octal number being sought.

Example: Convert the fractional quantity 0.42_{10} to its octal equivalent.

$$0.42 \times 8 = 3.36$$
$$.36 \times 8 = 2.88$$
$$.88 \times 8 = 7.04$$

Therefore $0.42_{10} = .327_8$.

Assignment: Using the method of repeated multiplication by 8, convert the fractional decimal numbers to octal equivalents. Carry out the conversion to obtain an octal number having one more significant digit than the decimal number being converted. Place the answers in the spaces provided.

_____ 1. $.625_{10}$
_____ 2. $.12_{10}$
_____ 3. $.75_{10}$
_____ 4. $.05_{10}$
_____ 5. $.25_{10}$

_____ 6. $.001_{10}$
_____ 7. $.999_{10}$
_____ 8. $.5_{10}$
_____ 9. $.8_{10}$
_____ 10. 214.32_{10}

Name_____ Date_____ Class_____ Grade_____

6. BINARY-TO-DECIMAL CONVERSION BY POSITIONAL VALUES

In the binary numbering system, the positional values are the powers of 2, e.g., 2^0, 2^1, 2^2, 2^3, 2^4, etc. These values are listed below.

$2^0 = 1$ $2^6 = 64$

$2^1 = 2$ $2^7 = 128$

$2^2 = 4$ $2^8 = 256$

$2^3 = 8$ $2^9 = 512$

$2^4 = 16$ $2^{10} = 1024$

$2^5 = 32$ etc.

Numbers expressed in the binary system can be converted to the decimal system by adding the positional values (powers of 2) corresponding to the bits (digits) of the binary number.

Example: Convert 10110_2 to N_{10}.

$$10110 = 1(2^4) + 0(2^3) + 1(2^2) + 1(2^1) + 0(2^0)$$
$$= 1(16) + 0(8) + 1(4) + 1(2) + 0(1)$$
$$= 16 + 4 + 2$$
$$= 22$$

Therefore $10110_2 = 22_{10}$.

This procedure may be used in reverse to convert from decimal to binary. The procedure is as follows: Subtract from the decimal number the largest power of 2 it contains, then subtract from the remainder the largest power of 2 it contains, etc. Continue until the remainder has been reduced to zero. The powers of 2 which have thus been subtracted indicate the positions of the 1's of the binary equivalent.

Example: Convert 75_{10} to its binary equivalent.

$$75 - 2^6 = 75 - 64 = 11$$
$$11 - 2^3 = 11 - 8 = 3$$
$$3 - 2^1 = 3 - 2 = 1$$
$$1 - 2^0 = 1 - 1 = 0$$

Therefore, $75_{10} = 2^6 + 2^3 + 2^1 + 2^0 = 1001011_2$.

Assignment: Perform the following conversions. Place answers in the spaces provided.

_____ 1. $101_2 = N_{10}$ _____ 6. $101101_2 = N_{10}$

_____ 2. $11001_2 = N_{10}$ _____ 7. $11111110_2 = N_{10}$

_____ 3. $1111_2 = N_{10}$ _____ 8. $100_{10} = N_2$

_____ 4. $10011110_2 = N_{10}$ _____ 9. $225_{10} = N_2$

_____ 5. $10101_2 = N_{10}$ _____ 10. $129_{10} = N_2$

```
:\CPM22\101LABS>
CSC101  Lab  8         Name/Date/Section  Dennis Barrett  / CSC101 / 12-10-90
```

Purpose: In this project you will write program code using mnemonic
          assembler operation codes and special subroutines.

1. Write your program to begin in memory location 100 and call the
   appropriate subroutine when needed. Put the subroutines into your
   program with the following command. DDT NEWTMPLT.DDT

2. Start your program by using the -a100<ret> command to initialize
   the starting code address. Suggestion: Use several (3-6) NOP codes
   at very beginning. This will allow you to add extra code without
   having to retype everything. When you need code to bring in INPUT
   or send OUTPUT, CALL <address> for the appropriate routine. See
   handout for descriptions and addresses of the following:

   KBD2ASC, ASC2CRT, KBD2HEX, BINIHEX, BINIDEC, CRLF, READA, WRITA/B

3. These routines are stored on the C disk in the 101LABS sub-
   directory. See file NEWTMPLT.DDT rev 3

4. Flow chart and write a program for one or more of the following.

   a. Write a program that issues instructions to input numbers, add
       the numbers together and then displays the answer and a
       message which identifies the answer.

   b. Write a program that issues instructions to input two numbers,
       subtracts the second number from the first, displays the
       difference and identifies the answer. (hardest)

   c. Write a program that prints a line of asterisks, skips a line,
       writes your name, course and date on one line, skips a line,
       writes a short message, of your own creation, skips a line, writes
       another line of asterisks and then prints "T H E   E N D", in
       the approximate middle of the next line. (easiest)

   d. Add any embellishments of your own with which you wish to
       experiment. Save program(s) as 101lb8a/b/c.cpm

5. List and run your program. Include printed program listing
   of your part of the program and runtime results. You optionally
   may include the listing of subroutines used.

                    PLEASE BE NEAT AND COMPLETE
Means - placing your data on the paper, stapling, full sheets, dating, etc.

Name_____ Date_____ Class_____ Grade_____

7. CONVERSION OF FRACTIONAL BINARY NOTATION TO DECIMAL SYSTEM

Two methods of converting fractional binary numbers to decimal equivalents are described below.

Method 1: Positional values of bits to the right of the binary point are 2^{-1}, 2^{-2}, 2^{-3}, etc. A fractional binary number therefore can be converted by adding the weights corresponding to the 1's in the binary number.

These values or weights are listed below.

$2^{-1} = 1/2$ $\qquad\qquad\qquad\qquad$ $2^{-4} = 1/16$

$2^{-2} = 1/4$ $\qquad\qquad\qquad\qquad$ $2^{-5} = 1/32$

$2^{-3} = 1/8$ $\qquad\qquad\qquad\qquad$ etc.

Example: Convert $.1101_2$ to N_{10}.

$$.1101_2 = 1(2^{-1}) + 1(2^{-2}) + 0(2^{-3}) + 1(2^{-4})$$
$$= \frac{1}{2} + \frac{1}{4} + \frac{1}{16}$$
$$= \frac{8}{16} + \frac{4}{16} + \frac{1}{16}$$
$$= \frac{13}{16} = .8125_{10}$$

Method 2: Move the binary point to the right of the least significant bit. Convert the resulting number to decimal notation and then divide by 2^x, where x is the number of places the binary point was moved.

Example: Convert $.1101_2$ to N_{10}.

$$.1101_2 = \frac{1101_2}{2^4} = \frac{13}{2^4} = \frac{13}{16} = .8125_{10}$$

The same methods can be used for converting fractional octal numbers except that in Method 1 use 8^{-1}, 8^{-2}, 8^{-3}, etc., and in Method 2 use 8^x instead of 2^x.

Assignment: Perform the following conversions. Place answers in the spaces provided.

_____ 1. $.101_2 = N_{10}$ $\qquad\qquad$ _____ 6. $.0101_2 = N_{10}$
_____ 2. $.001_2 = N_{10}$ $\qquad\qquad$ _____ 7. $.0011_2 = N_{10}$
_____ 3. $.1100_2 = N_{10}$ $\qquad\qquad$ _____ 8. $.526_8 = N_{10}$
_____ 4. $.111_2 = N_{10}$ $\qquad\qquad$ _____ 9. $.77_8 = N_{10}$
_____ 5. $.10001_2 = N_{10}$ $\qquad\qquad$ _____ 10. $.303_8 = N_{10}$

Name_____ Date_____ Class_____ Grade_____

8. DECIMAL-TO-BINARY CONVERSION BY REMAINDER METHOD

A number expressed in the decimal system can be converted to binary by successive divisions by 2. The remainder of each division is retained as a bit of the binary number, with the first remainder as the least significant bit.

Example: Convert 39_{10} to N_2.

$$\frac{39}{2} = 19 \text{ and remainder of } 1$$

$$\frac{19}{2} = 9 \text{ and remainder of } 1$$

$$\frac{9}{2} = 4 \text{ and remainder of } 1$$

$$\frac{4}{2} = 2 \text{ and remainder of } 0$$

$$\frac{2}{2} = 1 \text{ and remainder of } 0$$

$$\frac{1}{2} = 0 \text{ and remainder of } 1$$

Therefore, $39_{10} = 100111_2$.

Assignment: Using the remainder method, convert the following decimal system numbers to their binary equivalents. Place the answers in the spaces provided.

_____ 1. 25
_____ 2. 54
_____ 3. 432
_____ 4. 69
_____ 5. 128

_____ 6. 525
_____ 7. 274
_____ 8. 633
_____ 9. 1024
_____ 10. 1200

Name_____ Date_____ Class_____ Grade_____

9. CONVERSION OF FRACTIONAL DECIMAL NOTATION TO BINARY SYSTEM

Fractional quantities expressed in the decimal system can be converted to the binary system by repeated multiplication by 2. In the result of each multiplication, the digits to the right of the decimal point are used for the next multiplication. The digit to the left of the decimal point is retained as one of the bits of the binary number being sought, the first bit so obtained being the most significant bit.

Example: Convert 0.625_{10} to N_2.

$$.625 \times 2 = 1.250$$
$$.250 \times 2 = 0.500$$
$$.500 \times 2 = 1.000$$
$$.000 \times 2 = 0.000$$

Therefore, $0.625_{10} = .1010_2$.

In the above example, further multiplications by 2 would have produced only zeros in the answer. It is good practice however, to continue until the number of binary bits obtained is about three times the number of digits in the decimal number being converted. The decimal quantity .001, for example, is approximately equal to the binary quantity .0000000001.

Assignment: Using the method of repeated multiplication by 2, convert the following decimal system quantities to their binary equivalents.

_____ 1. .5
_____ 2. .333
_____ 3. .01
_____ 4. .99
_____ 5. .25
_____ 6. .167
_____ 7. .000001
_____ 8. .750
_____ 9. .1111
_____ 10. 25.25

10. BINARY-TO-DECIMAL CONVERSION BY DOUBLE-DABBLE METHOD

A method of converting binary numbers to decimal equivalents (referred to as the double-dabble or double-dibble method) is performed as follows. Write a 1 over the 1 farthest left in the binary number to be converted. Moving to the right, write a number over each bit according to this rule: if writing over a 0 bit, double the preceding number; if writing over a 1 bit, double the preceding number and add 1. The number written over the bit farthest right (the 2^0 position) is the decimal equivalent being sought.

Example: Convert 1000101_2 to N_{10}.

1	2	4	8	17	34	69
1	0	0	0	1	0	1

Therefore, $1000101_2 = 69_{10}$.

Assignment: Using the double-dabble method, convert the following binary numbers to decimal equivalents. Place answers in the spaces provided.

_____ 1. 100001

_____ 2. 11011

_____ 3. 111111

_____ 4. 1010101010

_____ 5. 1000

_____ 6. 101101101

_____ 7. 10011

_____ 8. 11110000

_____ 9. 100100

_____ 10. 1110011100

Name_____ Date_____ Class_____ Grade_____

11. OCTAL-BINARY CONVERSION BY SUBSTITUTION

A fast, convenient method of converting an octal number to its binary equivalent is to substitute a group of three binary bits for each octal digit. These substitutions are made according to the table shown below.

Example: Convert 705_8 to N_2. By substitution, $7 = 111$, $0 = 000$, and $5 = 101$. Therefore $705_8 = 111000101_2$.

Octal	Binary
0	000
1	001
2	010
3	011
4	100
5	101
6	110
7	111

For converting binary to octal notation, the procedure is reversed: Substitute an octal digit for each group of three binary bits. The groups of three are assumed to start from the binary point.

Example: Convert 11101000011_2 to N_8.

Group the bits into triplets: 011   101   000   011
Substitute for each group: 3   5   0   3
Therefore $11101000011_2 = 3503_8$.

Assignment: Perform the following conversions. Place answers in spaces provided.

_____ 1. $16_8 = N_2$
_____ 2. $11101_2 = N_8$
_____ 3. $1000000_2 = N_8$
_____ 4. $7777_8 = N_2$
_____ 5. $110010010010_2 = N_8$
_____ 6. $11011_2 = N_8$
_____ 7. $134.605_8 = N_2$
_____ 8. $1111_8 = N_2$
_____ 9. $110.111001_2 = N_8$
_____ 10. $10000_8 = N_2$

Name_____ Date_____ Class_____ Grade_____

12. BINARY-CODED DECIMAL NOTATION

Conversion from decimal notation to binary notation is generally a tedious procedure. It is much easier (for both man and machine) to perform the conversion on a digit-by-digit basis. Thus, for example, instead of converting 96_{10} to its *pure* binary equivalent, 1100000_2, it can be converted more easily digit-by-digit to 1001 0110. This form of notation is known as binary-coded decimal (BCD). It is also known as 8421 code because each decimal digit is represented by a group of four binary bits having weights (positional values) of 8, 4, 2 and 1. The four-bit representations for the decimal digits are shown in the table below.

Decimal	BCD
0	0000
1	0001
2	0010
3	0011
4	0100
5	0101
6	0110
7	0111
8	1000
9	1001

Assignment: Convert the following decimal numbers to BCD. Place the answers in the spaces provided.

_____ 1. 127_{10}

_____ 2. 99.76_{10}

_____ 3. 1200_{10}

_____ 4. 6.28_{10}

_____ 5. 13576428_{10}

Name_____ Date_____ Class_____ Grade_____

13. EXCESS-THREE CODE

The excess-three code permits conversion from the decimal system to binary form on a digit-by-digit basis. Conversion is accomplished by substituting a four-bit binary group for each decimal digit. The four-bit representations for the decimal digits are shown in the table below. Each representation is actually a binary number greater by 3_{10} than the decimal digit represented.

Example: Convert 732_{10} to excess-three code. By substitution 732 = 1010 0110 0101.

Decimal	XS3 Code	Decimal	XS3 Code
0	0011	5	1000
1	0100	6	1001
2	0101	7	1010
3	0110	8	1011
4	0111	9	1100

The excess-three code (abbreviated XS3) is a nonweighted code, i.e., it does not have positional values. An interesting characteristic of this code is that the *nines complement* can be obtained simply by changing the 1's to 0's and the 0's to 1's. As the table shows, any two digits whose sum is 9 (4 and 5, 6 and 3, etc.) are represented by four-bit groups that are opposites in terms of 1's and 0's. This feature of the code is advantageous in the design of certain types of arithmetic circuits for digital computers.

Assignment: Convert the following decimal numbers to XS3. Place answers in the spaces provided.

_____ 1. 209_{10}

_____ 2. 7463_{10}

_____ 3. 90954_{10}

_____ 4. 1001_{10}

_____ 5. 81_{10}

14. TWO-OUT-OF-FIVE CODE

The two-out-of-five code, like the 8421 and XS3 codes, permits conversion of decimal numbers to binary form on a digit-by-digit basis. As shown in the table, the two-out-of-five code (abbreviated 2/5) uses five-bit groups to represent the decimal digits.

Example: Convert 259_{10} to the 2/5 code. By substitution 259 = 00101 01100 11000.

Decimal	2/5 Code
0	00110
1	00011
2	00101
3	01001
4	01010
5	01100
6	10001
7	10010
8	10100
9	11000

An interesting feature of the 2/5 code is that every five-bit group contains two 1's (hence the name two-out-of-five). This characteristic makes the code self-checking, i.e., a machine employing this code can be designed to recognize errors and circuit failures which produce either more or fewer than two 1's per five-bit group.

Assignment: Using the above table for reference, convert the following decimal numbers to 2/5 code. Place answers in spaces provided.

1. 125_{10}
2. 3035_{10}
3. 100_{10}
4. 543_{10}
5. 63210_{10}

Name_____ Date_____ Class_____ Grade_____

15. MISCELLANEOUS CODES

The codes in the accompanying table are all employed on a digit-by-digit basis for converting from decimal to binary form. The 7421 code, so-called because these are the weights assigned to the four bits, requires no more than two 1's to represent any decimal digit.

The 2421 code is another example of a weighted four-bit code. In this code the *nines* complement can be obtained simply by changing 1's to 0's and 0's to 1's.

Seven-bit groups are employed in the biquinary code. This is a weighted code having positional values of 5, 0, 4, 3, 2, 1, and 0. The code is self-checking; that is, every legitimate seven-bit combination has two 1's. For this reason, it is sometimes called a two-out-of-seven code.

Decimal	7421 Code	2421 Code	Biquinary Code
0	0000	0000	0100001
1	0001	0001	0100010
2	0010	0010	0100100
3	0011	0011	0101000
4	0100	1010	0110000
5	0101	0101	1000001
6	0110	1100	1000010
7	1000	1101	1000100
8	1001	1110	1001000
9	1010	1111	1010000

Assignment: In problems 1 and 2, convert to 7421 code; in problems 3 and 4, convert to 2421 code; in problem 5, convert to biquinary code. Place answers in spaces provided.

_____ 1. 27096_{10}

_____ 2. 5009_{10}

_____ 3. 10377_{10}

_____ 4. 45.7_{10}

_____ 5. 749_{10}

16. ALPHANUMERIC CODES

Data processing machines employ binary codes to represent alphabetic information as well as numeric data. One such alphanumeric code is shown in the accompanying table. Each digit, alphabetic character, punctuation mark, etc. is represented by a unique six-bit combination. The digit 7, for example, is 000111; the letter X is 010111; the letter P is 100111; and the letter G is 110111. There are a total of 64 six-bit combinations, more than enough to represent the letters, digits, and some punctuation marks and special symbols.

In order to give the alphanumeric code an error-detecting characteristic, a seventh bit may be added to each six-bit group. This extra bit, usually placed in the position farthest left, is referred to either as a check bit or a parity bit. This bit is selected so that every seven-bit combination will have an odd number of 1's. The letter X, for example, would be 1010111. Computers employing such codes can be designed to recognize circuit failures which produce seven-bit combinations containing an *even* number of 1's. Some computers are designed for an even-parity check, i.e., every legitimate seven-bit combination has an even number of 1's.

	00	01	10	11
0000			−	+
0001	1	/	J	A
0010	2	S	K	B
0011	3	T	L	C
0100	4	U	M	D
0101	5	V	N	E
0110	6	W	O	F
0111	7	X	P	G
1000	8	Y	Q	H
1001	9	Z	R	I
1010	0		!	?
1011	#		$	.
1100		%	*	
1101				
1110				
1111				

Assignment: Using the accompanying table for reference, write the seven-bit combination (odd parity) for the following characters.

_____ 1. W   _____ 4. $

_____ 2. J   _____ 5. +

_____ 3. 8

Name_____ Date_____ Class_____ Grade_____

17. ADDITION OF BINARY NUMBERS

Because the binary system employs only two symbols, addition is a simple process. Several examples are shown below.

$$\begin{array}{cccc} 0 & 0 & 1 & 1 \\ +0 & +1 & +1 & 1 \\ \hline 0 & 1 & 10 & +1 \\ & & & \hline \\ & & & 11 \end{array}$$

Note that 1 plus 1 yields a sum of 0 and a carry of 1 into the next column, producing an answer of 10 (this is 10_2, not decimal 10). Similarly, 1 plus 1 plus 1 yields a sum of 1 and a carry of 1 into the next column.

Binary numbers, like decimal numbers, are added column by column from right to left. In the example shown below, note that carries are generated in the third, fifth, and sixth columns (from the right).

Example: 110110
           +110101
           ────────
            1101011

Assignment: Perform the following binary additions. Place answers in the spaces provided.

_____ 1.   1010
              +0101

_____ 2.   11011
              +00111

_____ 3.   111
              +111

_____ 4.   11001111
              +11010011

_____ 5.   111111
              +   1

_____ 6.   1010101
              +1010101

_____ 7.   10001
              +10001

_____ 8.   1011001
              +1111111

_____ 9.   1001
              +1000

_____ 10. 110.11
              + 01.11

39

18. BINARY SUBTRACTION BY DIRECT METHOD

Because the binary system employs only two symbols, there are only four basic subtractions.

```
   0       1       1       0
  −0      −1      −0      −1
  ──      ──      ──      ──
   0       0       1       1
```

Note that a *borrow* is required in the case of "1 from 0." This borrow is obtained from the next higher-order column (not shown in the above example). Since a borrow brings a 1 into the column, 0 minus 1 becomes 10 minus 1.

Binary numbers are subtracted column by column, progressing from right to left. Three examples are shown below.

```
                    0           0 1 1
   1 1 0 1         1̸0 1        1̸0̸0̸0 1
  −0 1 0 0        − 0 1 1       −     1 0
  ───────         ───────       ─────────
   1 0 0 1         0 1 0         0 1 1 1 1
```

The first example is straightforward; that is, no borrows are required. In the second example, subtraction in the second column requires a borrow from the third column (therefore changing the 1 to a 0 in the third column). The third example illustrates the case in which the required borrow cannot be obtained from the next higher-order column because the bit in this column is already a 0. Subtraction in the second column requires a borrow which must be obtained all the way from the fifth column. The zeros passed over are changed to 1's, and the 1 in the fifth column becomes a 0.

Assignment: Perform the following binary subtractions. Place answers in the space provided.

```
_____  1.   1110          _____  6.   10000
              −1010                         −00001

_____  2.    111           _____  7.   11111
              − 001                         −01110

_____  3.    101           _____  8.    110
              −  11                         − 011

_____  4.   11011          _____  9.   11110
              −10111                         −11010

_____  5.   10101          _____ 10.  1011.110
              −00110                        −1000.011
```

Name_____ Date_____ Class_____ Grade_____

19. COMPLEMENTS

Complements of numbers are useful in certain arithmetic operations in digital computers. There are two kinds of complements of interest here: the R complement (R for radix) and the R−1 (radix minus one) complement. In the decimal system, for example, there are *tens* complements and *nines* complements.

The R−1 complement of a number is determined by subtracting each digit from the highest value a single digit can assume in that number system. Thus, the R−1 complement in the decimal system is determined by subtracting each digit from 9.

Example: What is the R − 1 (nines) complement of 11345_{10}?

$$\begin{array}{r} 99999 \\ -11345 \\ \hline 88654 \end{array}$$

Therefore, 88654 is the nines complement of 11345.

As indicated by the above example, the procedure is so simple that the complement can be determined merely by inspection.

The R complement can most easily be found by first obtaining the R−1 complement and then adding a 1. Thus the tens complement of 6012_{10} is 3987 plus 1, or 3988.

In the binary system, the R−1 (ones) complement can be obtained simply by changing 1's to 0's and 0's to 1's.

Assignment: Determine the R−1 complement in problems 1 through 5, and the R complement in 6 through 10. Place answers in the spaces provided.

_____	1. 14273_{10}	_____	6. 264_{10}
_____	2. 50002_{10}	_____	7. 6300_{10}
_____	3. 101010_2	_____	8. 234_8
_____	4. 276_8	_____	9. 220_3
_____	5. 1202_3	_____	10. 11101_2

Name_____Date_____Class_____Grade_____

20. DECIMAL SUBTRACTION BY COMPLEMENT METHOD

The use of complements permits operations in subtraction to be converted to operations in addition. This is advantageous in digital computers because it permits the *adding* circuits to handle subtractions, reducing the total amount of circuitry required.

Subtraction by the complement method is performed by (1) converting the subtrahend to its $R-1$ complement, (2) adding instead of subtracting, and (3) adding the overflow (carry from the column farthest left) into the column farthest right. This last operation (step 3) is known as an *end-around carry*.

Example: Subtract 257_{10} from 624_{10} by the complement method.
Complement the subtrahend, then add:

$$\begin{array}{r} 624 \\ +\ 742 \\ \hline 1366 \end{array}$$

End-around carry:

$$\begin{array}{r} 366 \\ +\ \ \ 1 \\ \hline 367 \end{array}$$

Therefore, $624 - 257 = 367$.

A variation of the above procedure is to use the R complement instead of the $R-1$ complement (tens complement instead of nines complement). In this case, the overflow digit is dropped, i.e., no end-around carry is performed.

Assignment: Perform the following decimal subtractions by the complement method. Use the $R-1$ complement in problems 1 through 5, and the R complement in problems 6 through 10. Place answers in the spaces provided.

_____ 1. 216 − 154
_____ 2. 1550 − 1143
_____ 3. 999 − 635
_____ 4. 1206 − 1102
_____ 5. 999 − 725

_____ 6. 134 − 114
_____ 7. 50000 − 49999
_____ 8. 1632 − 1487
_____ 9. 84 − 24
_____ 10. 719 − 563

21. BINARY SUBTRACTION BY COMPLEMENT METHOD

Binary numbers can be subtracted by the complement method using either the *ones* or *twos* complement. In the latter case, the overflow is discarded (end-around carry is not performed).

Subtraction by the complement method assumes that the subtrahend and minuend have the same number and order of digits. For example, the binary subtraction 101101 − 1110 is assumed to mean 101101 − 001110. The leading 0's in the subtrahend must not be disregarded because they become 1's when the number is complemented.

Example:

$$\begin{array}{r}101101 \\ -001110\end{array} = \begin{array}{r}101101 \\ +110001 \\ \hline +011110\end{array} \text{(R} - 1 \text{ complement)}$$

$$\begin{array}{r}011110 \\ +1 \\ \hline 011111\end{array} \text{(end-around carry)}$$

A variation of procedure occurs when the subtrahend is larger than the minuend (subtracting a larger number from a smaller one). In such a case there will be no overflow. The answer (the sum of the minuend and the complemented subtrahend) must now be complemented to obtain the correct answer.

Assignment: Perform the following binary subtractions by the complement method. Use the R−1 (ones) complement in problems 1 through 5, and the R (twos) complement in problems 6 through 10.

_____ 1. 110101 − 100011
_____ 2. 1111 − 1010
_____ 3. 101111 − 1111
_____ 4. 101010 − 10101
_____ 5. 11011 − 11100

_____ 6. 11101 − 11001
_____ 7. 1011 − 0111
_____ 8. 1110111 − 1001
_____ 9. 101101 − 10010
_____ 10. 10111 − 11001

22. BINARY MULTIPLICATION AND DIVISION

The procedure for multiplying binary numbers is essentially the same as that for decimal multiplication. Each bit of the multiplier, starting with the least significant bit (the bit farthest to the right), is multiplied by the multiplicand. The partial products then are added to obtain the final product.

Example:

```
        1 0 0 1
      ×   1 0 1
        1 0 0 1
      0 0 0 0
    1 0 0 1
    1 0 1 1 0 1
```

The procedure for dividing binary numbers is also essentially the same as the decimal system procedure.

Example:

```
              1 1 1
      1 1 0 ) 1 0 1 0 1 0
              1 1 0
              1 0 0 1
                1 1 0
                  1 1 0
                  1 1 0
                      0
```

Assignment: Perform the following multiplications and divisions of binary numbers. Place answers in spaces provided.

_____ 1. 110 × 101
_____ 2. 111 × 111
_____ 3. 11101 × 1010
_____ 4. 11011 × 1011
_____ 5. 111.011 × 101.1

_____ 6. 10100/100
_____ 7. 111100/1100
_____ 8. 1111101/11001
_____ 9. 1101.1/10.01
_____ 10. 1100100/11

Name_____ Date_____ Class_____ Grade_____

23. OCTAL ADDITION AND SUBTRACTION

Addition of octal numbers is similar to decimal addition, i.e., add column from right to left, adding the carry (if any) to the next column to the left.

Example:

$$\begin{array}{r} 143_8 \\ +264_8 \\ \hline 427_8 \end{array}$$

Note that a carry is generated in the second column in the above example because $4_8 + 6_8 = 12_8$ (octal 12 is equivalent to decimal 10).

The procedure for octal subtraction is similar to that for decimal subtraction: Start with the column farthest right and work toward the left, borrowing when necessary from the column to the left.

Assignment: Complete the octal addition table below. Using this table for reference, perform the indicated octal additions. Place answers in the spaces provided.

Octal Addition Table

	0	1	2	3	4	5	6	7
0							6	
1								
2								
3		4						
4				7			12	
5								
6		10						
7								

_____ 1. 227_8
             $+140_8$

_____ 2. 10146_8
             $+33541_8$

_____ 3. 7777_8
             $+1_8$

_____ 4. 777_8
             $+777_8$

_____ 5. $156.77_8 + 1.01_8$

24. OCTAL MULTIPLICATION AND DIVISION

The procedure for octal multiplication is similar to that for decimal multiplication, that is, each digit of the multiplier is multiplied by the multiplicand, and the partial products are then added. Octal division is similar to decimal system division.

Example: Multiply 350_8 by 21_8.

$$\begin{array}{r} 350_8 \\ \times\ 21_8 \\ \hline 350 \\ 720 \\ \hline 7550_8 \end{array}$$

Example: Divide 7550_8 by 350_8.

$$\begin{array}{r} 21 \\ 350\,\overline{)7550} \\ 720 \\ \hline 350 \\ 350 \\ \hline 0 \end{array}$$

Assignment: Complete the octal multiplication table below. Using this table for reference, perform the indicated calculations. Place answers in spaces provided.

Octal Multiplication Table

	0	1	2	3	4	5	6	7
0								
1								
2					10			
3			6					
4							30	
5		5						
6								
7			16					

1. $31_8 \times 21_8$
2. $777_8 \times 7_8$
3. $2051_8 \times 1234_8$
4. $4466_8/22_8$
5. $100_8/50_8$

Name_____ Date_____ Class_____ Grade_____

25. HEXADECIMAL NUMBER SYSTEM

The hexadecimal number system has a radix of sixteen. While any sixteen symbols may be used, it is conventional to employ the ten decimal digits plus six letters of the alphabet: 0, 1, 2, 3, 4, 5, 6, 7, 8, 9, A, B, C, D, E, and F. Note that the decimal quantities 10, 11, 12, 13, 14, and 15 can be written as single digits in the hexadecimal system.

1. Continue this listing of successive integers to 02F.
    000
    001
    002
    003
    004
    005
    006
    007
    008
    009
    00A
    00B
    00C
    00D
    00E
    00F
    010

_____ 2. If this listing were continued, what number would come after FFF?

_____ 3. What is the weight of the third digit to the left of the hexadecimal point?

_____ 4. Write the number of degrees in a circle as a hexadecimal number.

_____ 5. Write the number of cents in a dollar as a hexadecimal number.

_____ 6. Write the number of cents in 16 dollars as a hexadecimal number.

_____ 7. Convert hexadecimal ABC to its decimal equivalent.

_____ 8. Convert hexadecimal 3072 to N_2.

_____ 9. Convert hexadecimal 3072 by writing a four-bit binary group for each hexadecimal digit. Compare the answer with that obtained in problem 8.

_____ 10. Convert hexadecimal F98C7 to N_2.

55

26. AND/OR RELATIONS

The AND relation is illustrated by the following electrical circuit. Examination of this circuit reveals that current from the battery must flow through *all* three pushbutton switches to reach light bulb L. The light is therefore lit only when A AND B AND C are actuated. This is expressed by the Boolean equation: $L = ABC$.

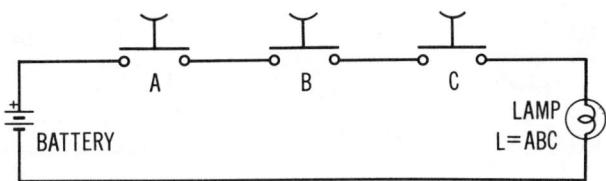

In the OR circuit shown below, current from the battery can reach the lamp if any one (or more) of the switches is actuated. The lamp therefore can be turned on by actuating A OR B OR C. This is expressed by the Boolean equation: $L = A + B + C$. In Boolean algebra, the $+$ means OR.

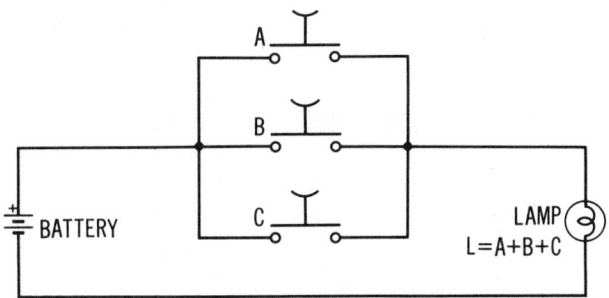

Assignment: Match the number of each figure to the number of the corresponding equation. Place answers in the table provided.

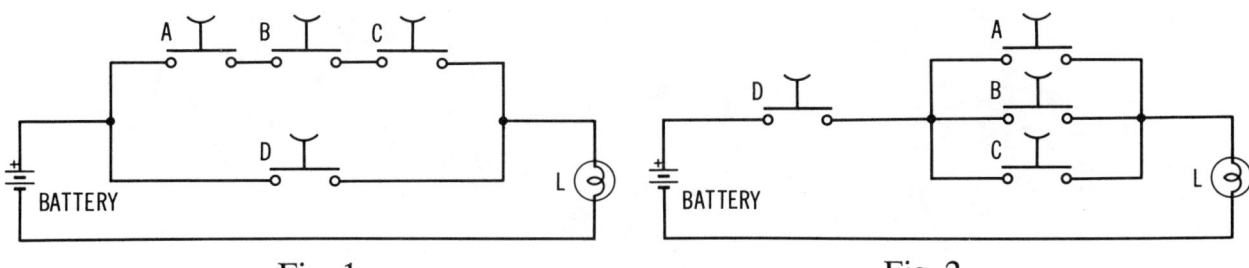

Fig. 1   Fig. 2

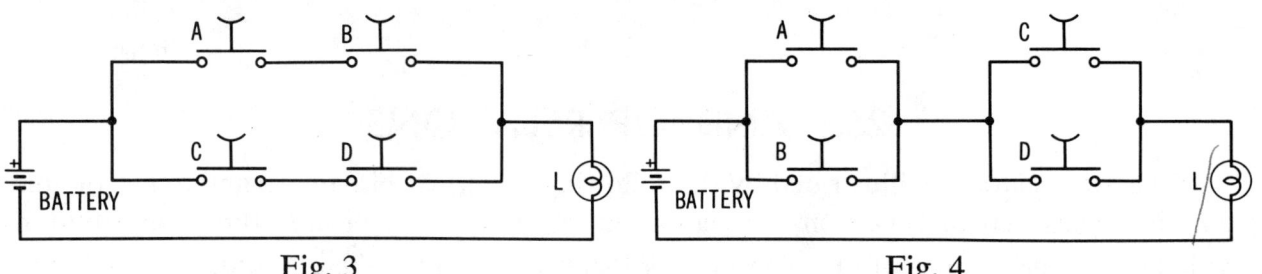

Fig. 3 Fig. 4

Equations:

I. L = D(A+B+C)
II. L = AB+CD
III. L = ABC+D
IV. L = (A+B) (C+D)

EQUATION	FIGURE
I	
II	
III	
IV	

27. AND/OR LOGIC

Digital computers employ high-speed electronic switches known as *logic* or *gate* circuits. In these circuits, two different voltage levels represent the binary symbols 0 and 1. One such circuit, the AND gate, is so designed that its output terminal will be at the voltage level which represents binary 1 when *all* of its input terminals are at the binary 1 level. If any input terminal is at the binary 0 level, the output will be binary 0. A block diagram of an AND gate is shown below with a table indicating the output for all combinations of input.

A	B	OUT
0	0	0
0	1	0
1	0	0
1	1	1

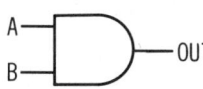

The AND gate represented above has two input terminals, A and B, but AND gates can be designed for a greater number of inputs. In every case, however, all inputs must be 1 to produce a 1 output.

An OR gate is a circuit designed so that its output terminal will be at the voltage level representing binary 1 when *at least* one of its input terminals is at the binary 1 level.

A	B	OUT
0	0	0
0	1	1
1	0	1
1	1	1

Assignment: Circle the Boolean equations which are correct for the logic diagram shown.

1. D = AB+C
2. D = ABC
3. G = E+F
4. H = ACE
5. H = AC+D
6. H = D+EF
7. G = EF
8. H = D+G
9. H = DG
10. H = EF+ABC

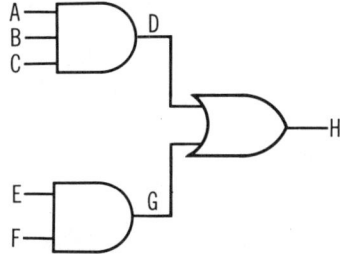

28. LOGIC SYMBOLS

Various symbols have been used to represent AND and OR gates. Although standards have been established, they have not been universally accepted. Some commonly employed symbols are shown below.

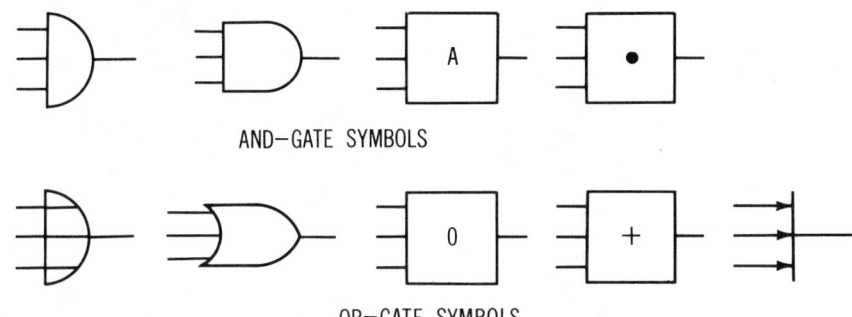

AND-GATE SYMBOLS

OR-GATE SYMBOLS

Variations of terminology also exist. The two voltage levels which represent binary 1 and binary 0 in electronic circuits are variously referred to as high and low, true and false, and logical one and logical zero.

The symbol ∨ is sometimes used instead of + to represent the OR relation ($A+B = A \vee B$), and the dot symbol or parentheses is sometimes used for the AND relation ($AB = A \cdot B = A(B)$). The terms *logical product* and *logical sum* are used in reference to the AND and OR functions respectively.

Assignment: Circle the equations which are correct for the logic diagram shown.

1. $G = A \cdot B \cdot C \cdot D$
2. $H = EF$
3. $J = G(HI)$
4. $H = EF$
5. $G = A \vee B \vee C \vee D$
6. $J = ABCDEF$
7. $H = E+F$

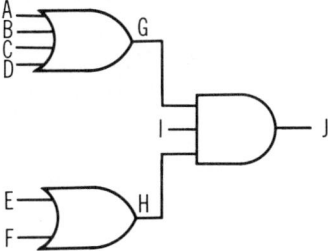

Draw logic diagrams corresponding to the following equations:

8. $L = MN+P$
9. $Z = (X \cdot Y)+(U \cdot V)$
10. $E = (A+B)(C+D)$

29. BOOLEAN POSTULATES

Several of the postulates of Boolean algebra are shown listed in the table below, for convenience.

$A+B = B+A$ $AB = BA$	Commutative Laws
$(AB)C = A(BC)$ $(A+B)+C = A+(B+C)$	Associative Laws
$A(B+C) = AB+AC$ $A+BC = (A+B)(A+C)$	Distributive Laws

The commutative laws, the associative laws, and the first of the two distributive laws compare directly to ordinary algebra. The second distributive law, however, is not so obvious. The validity of the relation $A+BC = (A+B)(A+C)$ is illustrated below with logic diagrams and corresponding tables. Comparison of the two tables reveals the equivalence of $A+BC$ and $(A+B)(A+C)$.

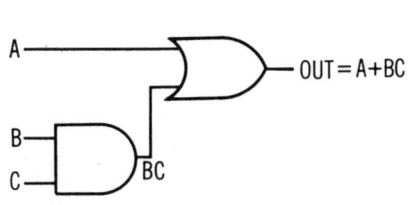
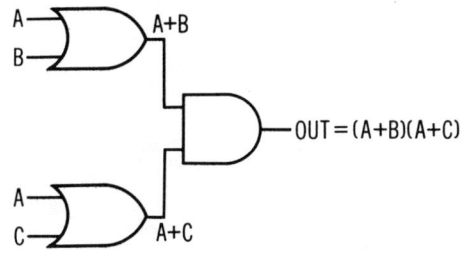

A	B	C	BC	A+BC
0	0	0	0	0
0	0	1	0	0
0	1	0	0	0
0	1	1	1	1
1	0	0	0	1
1	0	1	0	1
1	1	0	0	1
1	1	1	1	1

A	B	C	A+B	A+C	(A+B)(A+C)
0	0	0	0	0	0
0	0	1	0	1	0
0	1	0	1	0	0
0	1	1	1	1	1
1	0	0	1	1	1
1	0	1	1	1	1
1	1	0	1	1	1
1	1	1	1	1	1

Assignment: Match the number of each figure to the number of the corresponding equation. Place answers in the table.

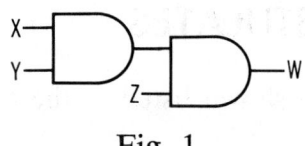

Fig. 1

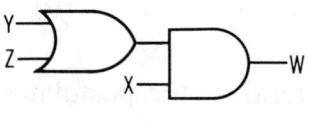

Fig. 2

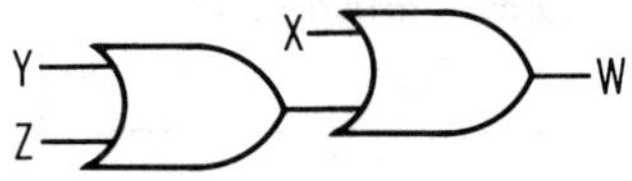

Fig. 3

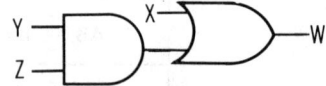

Fig. 4

Equations:

   I.   W = X+(Y+Z)
  II.   W = XYZ
 III.   W = (X+Y)(X+Z)
 IV.   W = X(Y+Z)

EQUATION	FIGURE
I	
II	
III	
IV	

30. THE NOT CONCEPT

The NOT concept is illustrated by the electrical circuit below. Because switch A has a normally closed contact, actuating the switch will break the circuit and turn off the lamp. The lamp is on only when the switch is NOT actuated. This is expressed by the Boolean equation $L = \overline{A}$ (sometimes written $L = A'$) which is read "L equals NOT A."

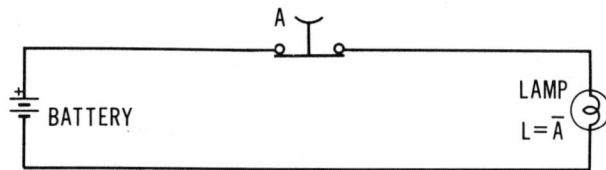

Electronic circuits which perform the NOT function are designed so that the output terminal is at the voltage level that represents binary 1 when the input terminal is at the binary 0 level. Also, the output level is binary 0 when the input level is binary 1. Such circuits are also known as *inverters*. Some commonly used schematic symbols are shown below.

Assignment: Match the number of the equation to the number of the corresponding diagram.

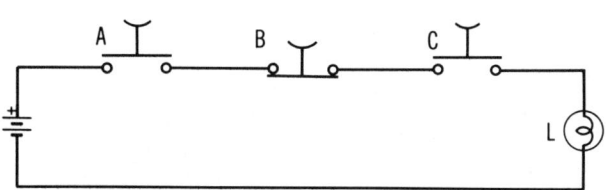

Fig. 1

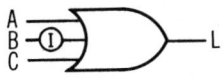 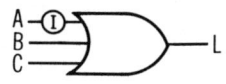 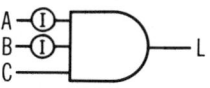 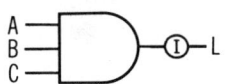

Fig. 2   Fig. 3   Fig. 4   Fig. 5

Equations:

  I. $L = \overline{A}BC$
 II. $L = \overline{A} + B + C$
III. $L = A\overline{B}C$
 IV. $L = A + \overline{B} + C$
  V. $L = \overline{ABC}$

EQUATION	FIGURE
I	4
II	2
III	1
IV	3
V	5

31. THEOREMS

Theorems of Boolean algebra, numbered for reference, are shown below. Also shown are block diagrams (figures) numbered to correspond to the theorems. Theorem 2, for example, is clarified by Fig. 2. Because an OR circuit will produce a 1 output when *at least* one of its inputs is logical 1, the output of Fig. 2 always will be a 1 (whether terminal A is 1 or 0). This constant output is represented by the 1 in the theorem: $A + \overline{A} = 1$.

Fig. 5 clarifies theorem 5. Since a constant 1 is applied to one of the terminals of the AND circuit, the output always will be the same as the input to terminal A. Hence $A \cdot 1 = A$.

(1) $A + A = A$
(2) $A + \overline{A} = 1$
(3) $A + 1 = 1$
(4) $A + 0 = A$
(5) $A \cdot 1 = A$
(6) $A \cdot 0 = 0$
(7) $A \cdot A = A$
(8) $A \cdot \overline{A} = 0$
(9) $1 + 0 = 1$
(10) $1 + 1 = 1$
(11) $1 \cdot 0 = 0$
(12) $1 \cdot 1 = 1$

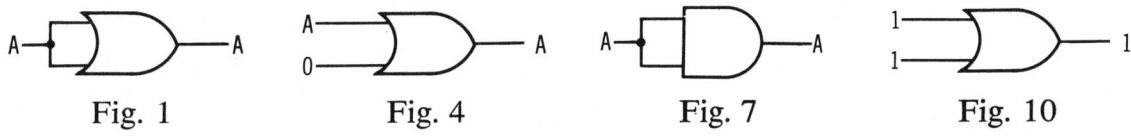

The above theorems often permit simplification of Boolean expressions.

Example: $B + \overline{B} + 0 = 1 + 0$ (by theorem 2)
$\qquad\qquad\qquad\quad\; = 1$ (by theorem 9)

Therefore, $B + \overline{B} + 0 = 1$.

Assignment: Using the theorems, reduce each expression to its simplest form. Place answers in the spaces provided.

_____ 1. 1 + 1 + 0
_____ 2. 1·1·A
_____ 3. M·$\overline{M}$·1
_____ 4. X·0 + 1
_____ 5. C·1 + $\overline{D}$D

_____ 6. A + 0 + A + 0
_____ 7. A + B + 1
_____ 8. 1(E + $\overline{E}$)
_____ 9. H + H + H + $\overline{H}$
_____ 10. 1·0·A

Name_____ Date_____ Class_____ Grade_____

32. REMOVING COMMON FACTORS

Boolean expressions and equations often can be simplified by removing common factors and applying the basic theorems (worksheet no. 31).

Example: Simplify the expression $A + AB$.

$A + AB = A(1 + B)$ (removing A as a common factor)
    $= A(1)$ (because $1 + B = 1$)
    $= A$ (because $A \cdot 1 = A$)

Example: Simplify the expression $AB + A\bar{B} + CDE\bar{E}$.

$AB + A\bar{B} + CDE\bar{E} = A(B + \bar{B}) + CDE\bar{E}$ (removing A from first two terms)
        $= A(1) + CDE\bar{E}$ (because $B + \bar{B} = 1$)
        $= A + CDE\bar{E}$ (because $A \cdot 1 = A$)
        $= A + CD \cdot 0$ (because $E\bar{E} = 0$)
        $= A + 0$ (because $CD \cdot 0 = 0$)
        $= A$ (because $A + 0 = A$)

Assignment: Reduce each expression to its simplest form. Place answers in the spaces provided.

_____ 1. $A + \bar{A} + B$ _____ 6. $EF + 0 + E\bar{F}$
_____ 2. $A + B + \bar{A} + AB$ _____ 7. $B + BC$
_____ 3. $AB + CDD + BD + 1$ _____ 8. $DE + DEF + DEG$
_____ 4. $AA + BC + 0$ _____ 9. $ABC + \bar{A}BC + BCD$
_____ 5. $A + B + A + B + C$ _____ 10. $\bar{A}(A + B) + C$

69

33. THE TRUTH TABLE

The variables of a Boolean expression can occur in either inverted or noninverted forms, e.g., $\overline{A}$ or A. The binary symbols 1 and 0 are employed to represent the two forms of the variable. Thus, if $A = 1$, then $\overline{A} = 0$.

A *truth table* is a tabulation of the value of a Boolean expression for all possible combinations of its variables. A truth table for the expression $AB + C$ is shown below. As there are three variables (A, B, and C) there are eight (2^3) possible combinations of these variables. The truth table therefore has eight rows. For each combination of truth values of variables, the value of the expression $AB + C$ is given in the right-hand column. Note that the value of the expression is 1 whenever both A AND B are 1, or whenever C is 1.

A truth table sometimes is drawn in graphical form as shown below.

TRUTH TABLE

A	B	C	AB+C
0	0	0	0
0	0	1	1
0	1	0	0
0	1	1	1
1	0	0	0
1	0	1	1
1	1	0	1
1	1	1	1

GRAPHICAL FORM OF TRUTH TABLE

Another truth table, for the expression $A\overline{B}C$, is shown below. Note that the value of the expression is 1 only when $A = 1$ AND $B = 0$ AND $C = 1$.

TRUTH TABLE

A	B	C	A$\overline{B}$C
0	0	0	0
0	0	1	0
0	1	0	0
0	1	1	0
1	0	0	0
1	0	1	1
1	1	0	0
1	1	1	0

GRAPHICAL FORM

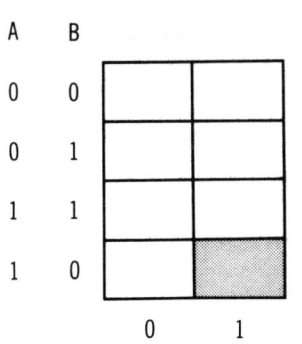

Assignment: Complete the following truth table for the expressions shown.

A	B	C	ABC	A + B + C	$\overline{A}\overline{B}\overline{C}$	$\overline{A}\overline{B}$ + C
0	0	0				
0	0	1				
0	1	0				
0	1	1				
1	0	0				
1	0	1				
1	1	0				
1	1	1				

34. CONSTRUCTION OF BOOLEAN TRUTH TABLES

When the Boolean expression is complex, it is sometimes best to construct the truth table for part of the expression at a time, and then combine the parts. An example, for the expression AB + BC + AC, is shown below. A separate column is provided for each of the three parts of the expression. The column labeled AB has a 1 when A AND B are both 1; the BC column has a 1 when B AND C are both 1; and the AC column shows a 1 for each occurrence of A AND C. These three columns then are combined to form the last column. The last column has a 1 when there is a 1 in the AB column OR the BC column OR the AC column.

A	B	C	AB	BC	AC	AB + BC + AC
0	0	0	0	0	0	0
0	0	1	0	0	0	0
0	1	0	0	0	0	0
0	1	1	0	1	0	1
1	0	0	0	0	0	0
1	0	1	0	0	1	1
1	1	0	1	0	0	1
1	1	1	1	1	1	1

Assignment: Complete the following truth tables.

Truth Table 1

A	B	C	AC	$\bar{B}$	AC + $\bar{B}$
0	0	0			
0	0	1			
0	1	0			
0	1	1			
1	0	0			
1	0	1			
1	1	0			
1	1	1			

Truth Table 2

A	B	C	$\bar{A}$	$\bar{A}B$	$\bar{A}B + C$
0	0	0			
0	0	1			
0	1	0			
0	1	1			
1	0	0			
1	0	1			
1	1	0			
1	1	1			

Truth Table 3

A	B	C	$\bar{B}$	$A + \bar{B} + C$
0	0	0		
0	0	1		
0	1	0		
0	1	1		
1	0	0		
1	0	1		
1	1	0		
1	1	1		

Truth Table 4

A	B	C	$\bar{B}$	$\bar{C}$	$A + \bar{B}\bar{C}$
0	0	0			
0	0	1			
0	1	0			
0	1	1			
1	0	0			
1	0	1			
1	1	0			
1	1	1			

Name_____ Date_____ Class_____ Grade_____

35. USE OF TRUTH TABLE

The truth table is useful for proving the equivalence (or lack of equivalence) of Boolean expressions. The following truth table, for example, proves the equivalence: $A + AB = A$. Column AB of the table contains a 1 when A and B are both 1. Column $A + AB$ contains a 1 when there is a 1 in either column A or column AB. A comparison of column $A + AB$ and column A reveals their equivalence.

A	B	AB	A+AB
0	0	0	0
0	1	0	0
1	0	0	1
1	1	1	1

Assignment: Complete the following truth tables to prove the equivalence of the expressions shown.

Truth Table 1

A	B	A+B	A(A+B)
0	0		
0	1		
1	0		
1	1		

$A(A+B) = A$

Truth Table 2

A	B	$\bar{A}$	$\bar{A}B$	$A+\bar{A}B$	A+B
0	0				
0	1				
1	0				
1	1				

$A + \bar{A}B = A + B$

Truth Table 3

A	B	C	B+C	A(B+C)	AB	AC	AB+AC
0	0	0					
0	0	1					
0	1	0					
0	1	1					
1	0	0					
1	0	1					
1	1	0					
1	1	1					

$A(B+C) = AB + AC$

Truth Table 4

A	B	C	BC	A+BC	A+B	A+C	(A+B)(A+C)
0	0	0					
0	0	1					
0	1	0					
0	1	1					
1	0	0					
1	0	1					
1	1	0					
1	1	1					

$A + BC = (A+B)(A+C)$

Name____________________Date__________Class______Grade______

36. CONVERTING BOOLEAN EQUATIONS TO BLOCK DIAGRAMS

Assignment: In the spaces provided, draw a block diagram corresponding to each equation. Label all terminals. Assume that the variables are available in noninverted form only, i.e., if an inverted form of the variable is required, a NOT circuit must be used.

1. $X = A\bar{B}$
2. $Y = \bar{A}BC + A\bar{B}\bar{C}$
3. $Z = A + B + C$
4. $X = \bar{A} + B$
5. $Z = A\bar{B} + \bar{A}B$

Fig. 1

Fig. 2

Fig. 3

Fig. 4

Fig. 5

37. CONVERTING BLOCK DIAGRAMS TO BOOLEAN EQUATIONS

Assignment: In the spaces provided, write the Boolean equation corresponding to each block diagram.

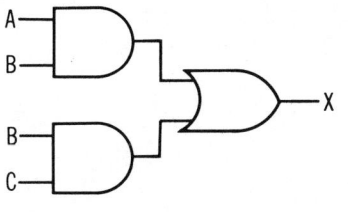

Fig. 1

X =

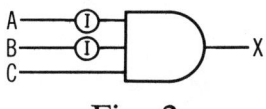

Fig. 2

X =

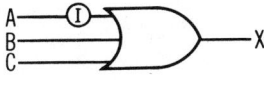

Fig. 3

X =

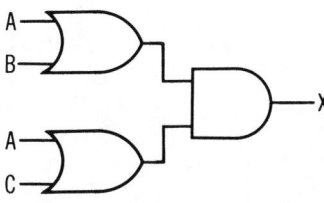

Fig. 4

X =

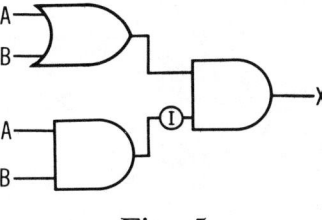

Fig. 5

X =

38. CONVERTING BLOCK DIAGRAMS TO TRUTH TABLES

Assignment: Complete the truth table for each block diagram.

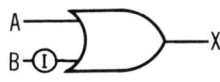

Fig. 1

Fig. 2

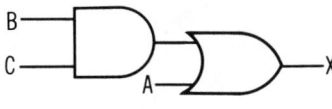

Fig. 3

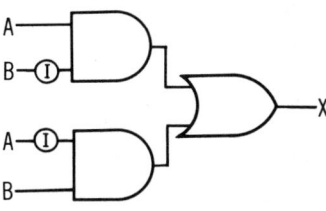

Fig. 4

Table 1

A	B	X
0	0	
0	1	
1	0	
1	1	

Table 2

A	B	X
0	0	
0	1	
1	0	
1	1	

Table 3

A	B	C	X
0	0	0	
0	0	1	
0	1	0	
0	1	1	
1	0	0	
1	0	1	
1	1	0	
1	1	1	

Table 4

A	B	X
0	0	
0	1	
1	0	
1	1	

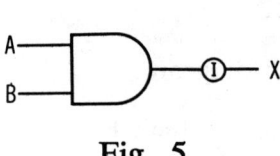

Fig. 5

Table 5

A	B	X
0	0	
0	1	
1	0	
1	1	

39. CONVERTING TRUTH TABLES TO BLOCK DIAGRAMS

Assignment: In the spaces provided, draw a block diagram corresponding to each table. Label all terminals. Assume that the variables are available in noninverted form only, i.e., if an inverted form of a variable is required, a NOT circuit must be used.

Table 1

A	B	C	X
0	0	0	0
0	0	1	0
0	1	0	0
0	1	1	1
1	0	0	0
1	0	1	1
1	1	0	1
1	1	1	1

Fig. 1

Table 2

A	B	C	X
0	0	0	0
0	0	1	1
0	1	0	1
0	1	1	1
1	0	0	1
1	0	1	1
1	1	0	1
1	1	1	1

Fig. 2

Table 3

A	B	C	X
0	0	0	0
0	0	1	1
0	1	0	1
0	1	1	0
1	0	0	1
1	0	1	0
1	1	0	0
1	1	1	0

Fig. 3

Table 4

A	B	C	X
0	0	0	0
0	0	1	0
0	1	0	0
0	1	1	1
1	0	0	0
1	0	1	1
1	1	0	1
1	1	1	0

Fig. 4

40. CONVERTING BOOLEAN EQUATIONS TO TRUTH TABLES

Assignment: Complete the truth table corresponding to each Boolean equation.

1. $X = \overline{A}\overline{B}$

   Table 1

A	B	X
0	0	
0	1	
1	0	
1	1	

2. $X = A\overline{B} + \overline{A}B + AB$

   Table 2

A	B	X
0	0	
0	1	
1	0	
1	1	

3. $X = AB + BC$

   Table 3

A	B	C	X
0	0	0	
0	0	1	
0	1	0	
0	1	1	
1	0	0	
1	0	1	
1	1	0	
1	1	1	

4. $X = AB + \overline{A}\overline{B}$

   Table 4

A	B	X
0	0	
0	1	
1	0	
1	1	

5. $X = A\bar{B} + \bar{A}B$

Table 5

A	B	X
0	0	
0	1	
1	0	
1	1	

41. CONVERTING TRUTH TABLES TO BOOLEAN EQUATIONS

Assignment: In the spaces provided, write the Boolean equation corresponding to each truth table.

Table 1

A	B	X
0	0	0
0	1	1
1	0	1
1	1	1

Eq. 1. $X =$

Table 2

A	B	X
0	0	0
0	1	0
1	0	1
1	1	0

Eq. 2. $X =$

Table 3

A	B	X
0	0	0
0	1	1
1	0	1
1	1	0

Eq. 3. $X =$

Table 4

A	B	X
0	0	1
0	1	1
1	0	1
1	1	0

Eq. 4. $X =$

Table 5

A	B	X
0	0	1
0	1	0
1	0	0
1	1	0

Eq. 5. $X =$

Name_____ Date_____ Class_____ Grade_____

42. THE VENN DIAGRAM

The Venn diagram is a graphical representation of a logical relation. The diagram is drawn as a set of circles enclosed by a rectangle—one circle for each variable. Although the Venn diagram does not contain any more information than the corresponding Boolean expression, it presents the information in a form that is easy to visualize. Venn diagrams are useful for representing relations involving up to four variables. Beyond that, the diagram becomes too complex and loses its effectiveness as a means of clarifying logical relations.

Several Venn diagrams are shown below. Note that the shaded part of each diagram corresponds to the Boolean expression it represents. For example, in the diagram corresponding to $X = AB$, the shaded area is in both A AND B. In the diagram corresponding to $X = A + B$, the shaded area is in A OR in B. In the diagram for $X = A\bar{B}C$, the shaded area is in A AND C but NOT in B. In the diagram representing $X = \bar{A} + B$, the shaded area is either (1) NOT in A OR (2) in B. In the diagram for $X = \bar{A}$, all the shaded area is NOT in A.

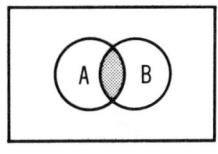 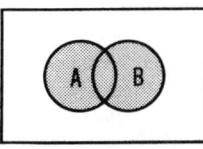 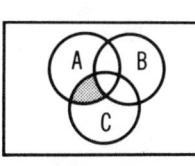 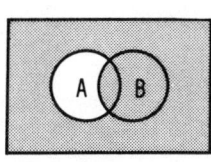 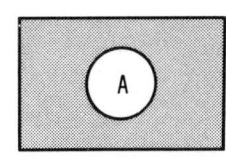

   X=AB   X=A+B   X=A$\bar{B}$C   X=$\bar{A}$+B   X=$\bar{A}$

Assignment: Complete the Venn diagrams for the Boolean expressions shown.

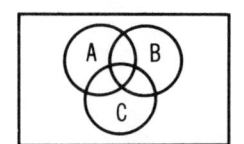

Fig. 1.   X = ABC

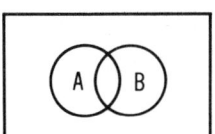

Fig. 2.   X = A$\bar{B}$

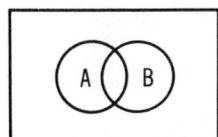

Fig. 3.   X = $\overline{AB}$

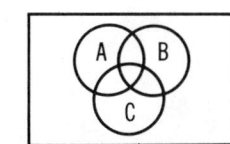

Fig. 4.   X = $\bar{A}$ + BC

43. CONVERTING VENN DIAGRAMS TO BOOLEAN EXPRESSIONS

Assignment: In the spaces provided, write the Boolean expressions corresponding to the Venn diagrams.

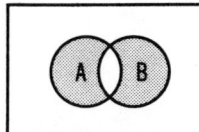

Fig. 1

X =

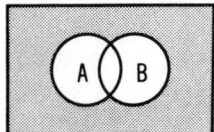

Fig. 2

X =

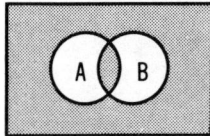

Fig. 3

X =

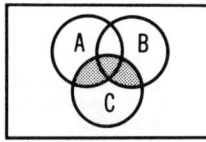

Fig. 4

X =

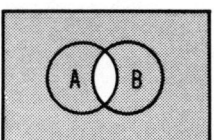

Fig. 5

X =

44. CONVERTING VENN DIAGRAMS TO BLOCK DIAGRAMS

Assignment: Draw a block diagram corresponding to each Venn diagram.

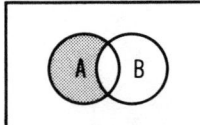

Fig. 1

Diagram 1

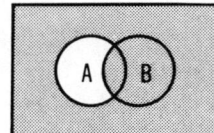

Fig. 2

Diagram 2

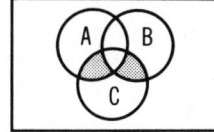

Fig. 3

Diagram 3

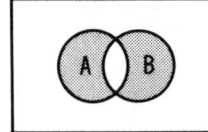

Fig. 4

Diagram 4

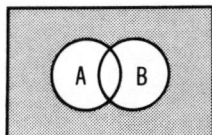

Fig. 5

Diagram 5

45. CONVERTING BLOCK DIAGRAMS TO VENN DIAGRAMS

Assignment: Complete the Venn diagrams corresponding to each block diagram.

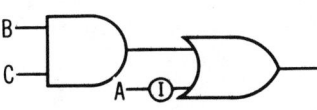

Fig. 1

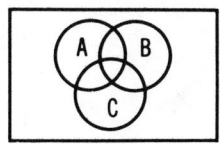

Diagram 1

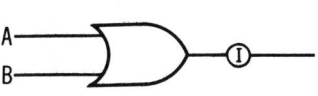

Fig. 2

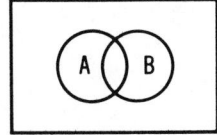

Diagram 2

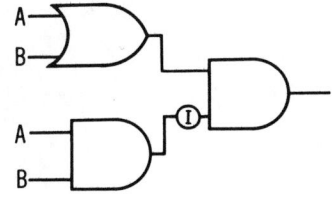

Fig. 3

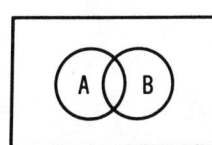

Diagram 3

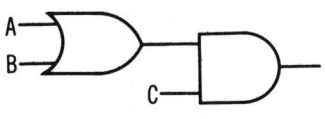

Fig. 4

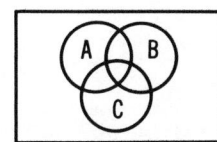

Diagram 4

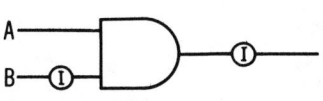

Fig. 5

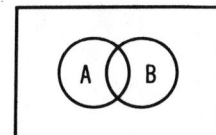

Diagram 5

46. REPETITION OF BOOLEAN EXPRESSIONS

Any of the terms of a Boolean expression or equation can be repeated any number of times without changing the logical meaning of the expression or equation. Thus, for example, $XYZ = XXYYZZZZ$ and $ABC + \bar{A}B\bar{C} + A\bar{B}C = ABC + \bar{A}B\bar{C} + A\bar{B}C + A\bar{B}C + \bar{A}B\bar{C}$. This rule is an extension of theorems $AA = A$ and $A + A = A$.

It is sometimes convenient to repeat a term of a Boolean equation so that it can be used with more than one group for removing common factors. For example, the expression $\bar{A}BC + ABC + AB\bar{C}$ can be simplified as follows:

$$\bar{A}BC + ABC + AB\bar{C} = \bar{A}BC + ABC + AB\bar{C} + ABC \quad \text{(repeating ABC)}$$
$$= BC(\bar{A} + A) + AB(\bar{C} + C) \quad \text{(removing common factors)}$$
$$= BC(1) + AB(1)$$
$$= BC + AB$$
$$= B(C + A) \quad \text{(removing B as common factor)}$$

Assignment:

1. Using the method of repetition, prove that $A\bar{B} + AB + \bar{A}B = A + B$. (Hint: Start with the expression $A\bar{B} + AB + \bar{A}B$, repeat the term AB, then simplify.)

2. Complete the two Venn diagrams to show that $A\bar{B} + AB + \bar{A}B = A + B$.

3. Complete the tables to show the equivalence of the two block diagrams.

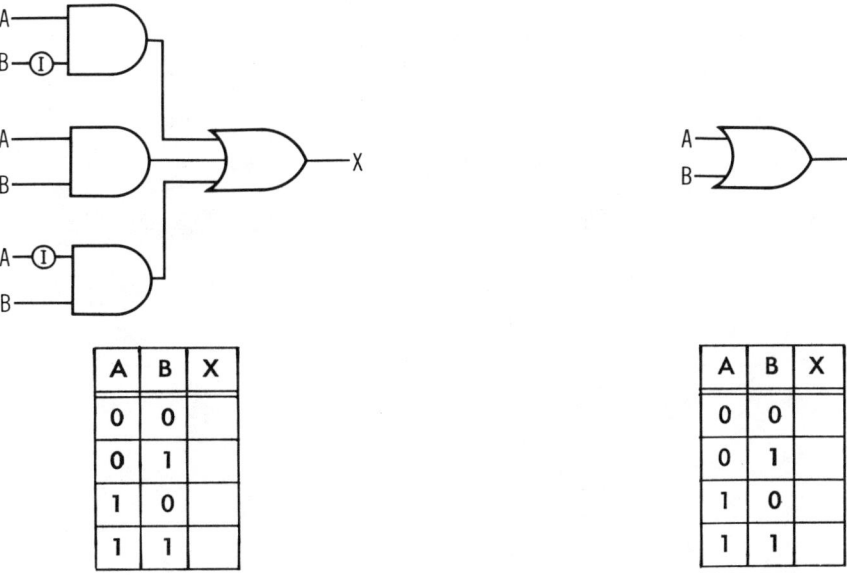

A	B	X
0	0	
0	1	
1	0	
1	1	

A	B	X
0	0	
0	1	
1	0	
1	1	

47. SOME USEFUL RELATIONS

Simplification of Boolean equations often can be accomplished by substitution of a simpler expression for a more complex one having the same logical meaning. For example, if an expression of the form $A\bar{B} + AB + \bar{A}B$ appears in a Boolean equation, it can be removed and replaced by $A + B$. This relation and several others are listed below.

(1) $A\bar{B} + AB + \bar{A}B = A + B$ (4) $A(A + B) = A$

(2) $AB + \bar{A}B + A\bar{B} + \bar{A}\bar{B} = 1$ (5) $A(\bar{A} + B) = AB$

(3) $A + \bar{A}B = A + B$ (6) $A + AB = A$

Example: Simplify $ABC + A\bar{B}C + AB\bar{C} + A\bar{B}\bar{C} + \bar{A}BC$.

$ABC + A\bar{B}C + AB\bar{C} + A\bar{B}\bar{C} + \bar{A}BC = A(BC + \bar{B}C + B\bar{C} + \bar{B}\bar{C}) + \bar{A}BC$
$\phantom{ABC + A\bar{B}C + AB\bar{C} + A\bar{B}\bar{C} + \bar{A}BC}$ (removing common factor)
$= A(1) + \bar{A}BC$ (relation 2 above)
$= A + \bar{A}BC$
$= A + BC$ (relation 3 above)

Assignment: Reduce each of the following to its simplest form. Place answers in the spaces provided.

_____ 1. $A + AB + A(A + B)$

_____ 2. $B(A + \bar{A}B)$

_____ 3. $A + AB + \bar{A}B + A\bar{B} + \bar{A}\bar{B} + AB$

_____ 4. $\bar{A}BC + A\bar{B}C + ABC$

_____ 5. $C(\bar{C} + AB)$

99

48. SIMPLIFICATION OF LOGIC DIAGRAMS BY USE OF BASIC THEOREMS

By the use of Boolean algebra, the practitioner of the electronics art can often effect savings in circuitry and cost. The block diagram shown below, for example, can be considerably reduced without altering its logical behavior.

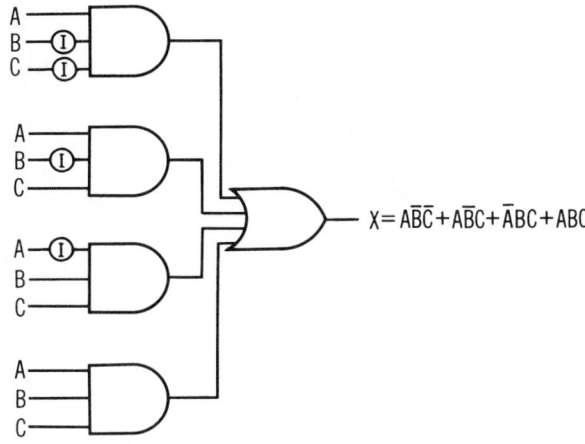

The equation for the output terminal (X) can be simplified:

$$X = A\bar{B}\bar{C} + A\bar{B}C + \bar{A}BC + ABC$$
$$= A\bar{B}(\bar{C} + C) + BC(\bar{A} + A)$$
$$= A\bar{B} + BC$$

A block diagram corresponding to the simplified equation can now be drawn:

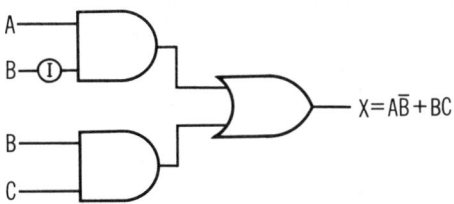

This block diagram is the logical equivalent of the more complex diagram from which it was derived. This equivalence can be proved by constructing a truth table or a Venn diagram for both block diagrams.

Assignment: Simplify each block diagram by reducing its output equation to its simplest form and then drawing a corresponding block diagram. Place the simplified diagrams in the spaces provided.

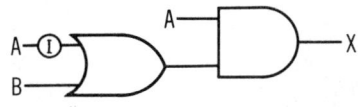

Fig. 1. $X = A(\bar{A} + B)$ Fig. 1

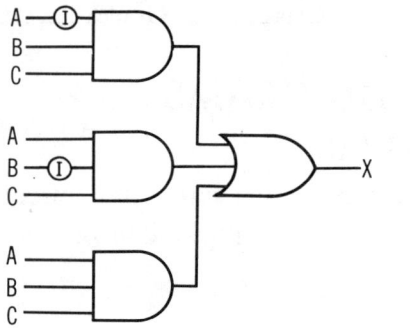

Fig. 2. $X = \bar{A}BC + A\bar{B}C + ABC$ Fig. 2

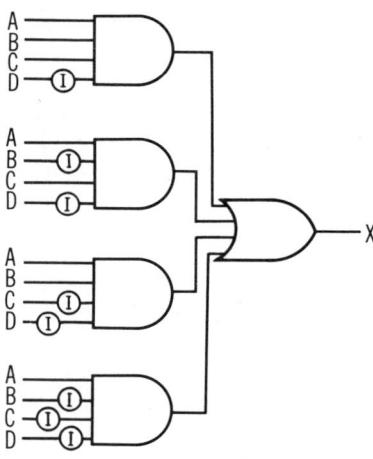

Fig. 3. $X = ABC\bar{D} + \bar{A}BCD + AB\bar{C}D + \bar{A}\bar{B}\bar{C}\bar{D}$ Fig. 3

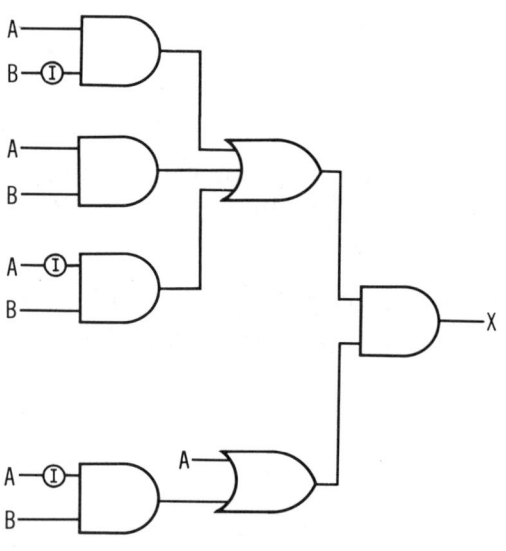

Fig. 4. $X = (A\bar{B} + AB + \bar{A}B)(A + \bar{A}B)$ Fig. 4

49. $\overline{AB} \neq \overline{A}\overline{B}$

It is important to note that an expression with a single NOT bar extending over several variables is not equal to the same expression with individual NOT bars over the variables. Thus $\overline{AB}$ is not the same as $\overline{A}\overline{B}$. That they are not equal is apparent from the truth table below.

A	B	AB	$\overline{AB}$	$\overline{A}\overline{B}$
0	0	0	1	1
0	1	0	1	0
1	0	0	1	0
1	1	1	0	0

A 1 appears in the AB column only when A AND B are both 1. Because $\overline{AB}$ is the exact opposite (negation) of AB, a 1 appears in the $\overline{AB}$ column wherever there is a 0 in the AB column, i.e., if AB = 1, then $\overline{AB}$ = 0. The $\overline{A}\overline{B}$ column shows a 1 only when A = 0 AND B = 0. Comparison of the $\overline{AB}$ and the $\overline{A}\overline{B}$ columns reveals their lack of equivalence.

That $\overline{AB}$ is not the same as $\overline{A}\overline{B}$ is further illustrated by the Venn diagrams below. Note that the diagram for $\overline{AB}$ is shaded in the area that is NOT in "A AND B." The diagram for $\overline{A}\overline{B}$ is shaded in the area that is "NOT in A" AND "NOT in B."

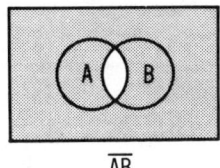
$\overline{AB}$

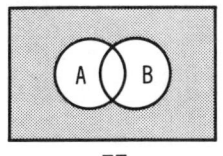
$\overline{A}\overline{B}$

Assignment: Complete the truth table and the Venn diagrams to show that $\overline{A + B}$ is not the same as $\overline{A} + \overline{B}$.

A	B	A+B	$\overline{A+B}$	$\overline{A}$	$\overline{B}$	$\overline{A} + \overline{B}$
0	0					
0	1					
1	0					
1	1					

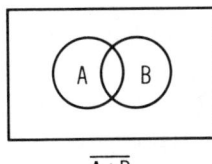
$\overline{A+B}$

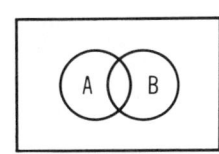
$\overline{A}+\overline{B}$

50. DE MORGAN'S THEOREM

In the truth table below, the $\bar{A}$ column is the exact opposite of the A column, i.e., if $A = 0$, then $\bar{A} = 1$, and vice versa. Similarly, the $\bar{B}$ column is the opposite of the B column.

A	B	$\bar{A}$	$\bar{B}$	$\bar{A} + \bar{B}$	AB	$\overline{AB}$
0	0	1	1	1	0	1
0	1	1	0	1	0	1
1	0	0	1	1	0	1
1	1	0	0	0	1	0

A 1 appears in the $\bar{A} + \bar{B}$ column whenever there is a 1 in the $\bar{A}$ column OR in the $\bar{B}$ column. Column $\overline{AB}$ is the exact opposite of column AB. Column $\bar{A} + \bar{B}$ and column $\overline{AB}$ are identical. Therefore: $\overline{AB} = \bar{A} + \bar{B}$. This relation is known as De Morgan's theorem, sometimes referred to as the *law of duality*.

It also can be shown that $\overline{A + B} = \bar{A}\bar{B}$. For example, if you are told to proceed on a course that is NOT "left or right" ($\overline{L + R}$), you would go straight ahead. Similarly, if you are told to proceed on a course that is "NOT left" AND "NOT right" ($\bar{L}\bar{R}$), you also would proceed straight ahead. Therefore, $\overline{L + R} = \bar{L}\bar{R}$. This relation also is part of De Morgan's theorem, rewritten below for reference:

$$\overline{ABC \cdots} = \bar{A} + \bar{B} + \bar{C} + \cdots$$

also: $\overline{A + B + C + \cdots} = \bar{A}\bar{B}\bar{C} \cdots$

Assignment: Complete the following truth table to show that $\overline{A + B + C} = \bar{A}\bar{B}\bar{C}$.

A	B	C	A+B+C	$\overline{A+B+C}$	$\bar{A}$	$\bar{B}$	$\bar{C}$	$\bar{A}\bar{B}\bar{C}$
0	0	0	0	1	1	1	1	1
0	0	1	1	0	1	1	0	0
0	1	0	1	0	1	0	1	0
0	1	1	1	0	1	0	0	0
1	0	0	1	0	0	1	1	0
1	0	1	1	0	0	1	0	0
1	1	0	1	0	0	0	1	0
1	1	1	1	0	0	0	0	0

51. APPLICATION OF DE MORGAN'S THEOREM

De Morgan's theorem permits a single NOT bar over several variables to be split into individual NOT bars, but the AND relations must be changed to OR, and vice versa. Also, individual NOT bars can be combined into a single NOT bar, but the AND's must be changed to OR's, and vice versa. Of course, parentheses or other signs of grouping must be considered.

Example: $\overline{(A+B)C} = \overline{(A+B)} + \overline{C}$
$= \overline{A}\overline{B} + \overline{C}$

Assignment: By applying De Morgan's theorem, rewrite each of the following expressions so that the single NOT bar over several variables is split into individual NOT bars. Place answers in spaces provided.

1. $\overline{ABCD}$
2. $\overline{AB + CD}$
3. $\overline{(A+C)\overline{B}}$
4. $\overline{ABC + D + E}$
5. $\overline{\overline{ABCD}}$

Rewrite the following expressions so that the individual NOT bars are combined into a single NOT bar (continue until there is only one NOT bar).

6. $\overline{A} + \overline{B} + \overline{C} + \overline{D}$
7. $\overline{A}\overline{B}\overline{C} + \overline{D}\overline{E}$
8. $(\overline{A} + \overline{C})\,\overline{B}$
9. $\overline{A}\overline{B}\overline{C} + \overline{D} + \overline{E}$
10. $\overline{A}\overline{B}\overline{C}\overline{D}$

52. TRUE-OUTPUT AND FALSE-OUTPUT FORMS

A Boolean equation can be written in either true-output form or false-output form. For example, consider the block diagram below.

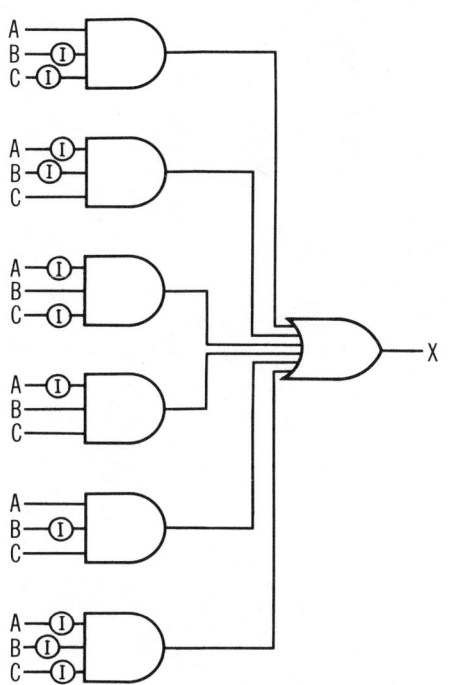

$$X = AB\bar{C} + \bar{A}BC + \bar{A}B\bar{C} + \bar{A}BC + A\bar{B}C + \bar{A}\bar{B}\bar{C}$$

A	B	C	X
0	0	0	1
0	0	1	1
0	1	0	1
0	1	1	1
1	0	0	1
1	0	1	1
1	1	0	0
1	1	1	0

Because there are three variables (A, B, and C) there are eight (2^3) possible combinations of these variables. Six of these combinations, indicated by the equation and the table, will produce a *true* output (X = 1) at terminal X. The two remaining combinations, not indicated in the equation, will produce a *false* output (X = 0) at X. The false-output form of the equation therefore can be written:

$$\bar{X} = AB\bar{C} + ABC$$

This equation is obviously easier to work with than the true-output form, although the two forms are logically equivalent. The false-output equation can be simplified as follows:

$$\bar{X} = AB\bar{C} + ABC$$
$$\bar{X} = AB(\bar{C} + C)$$
$$\bar{X} = AB$$

This simplified equation indicates that terminal X must be false (X = 0) when A AND B are both true (A = 1 AND B = 1). The corresponding diagram now can be drawn:

109

This diagram is logically equivalent to the more complex diagram from which it was derived.

Assignment: Write the false-output form of the following equations.

1. $X = A\bar{B} + \bar{A}B$ 1. $\bar{X} = (\bar{A}+B)(A+\bar{B})$

2. $X = \bar{A}B + A\bar{B} + AB$ 2. $\bar{X} = (A+\bar{B})(\bar{A}+B)(\bar{A}+\bar{B})$

3. $X = \bar{A}BC + \bar{A}B\bar{C} + \bar{A}\bar{B}\bar{C} + AB\bar{C}$ 3. $\bar{X} = (A+\bar{B}+\bar{C})(A+\bar{B}+C)(A+B+C)(\bar{A}+\bar{B}+C)$

4. $X = \bar{A}BC + A\bar{B}C + ABC + AB\bar{C} + \bar{A}\bar{B}\bar{C}$ 4. $\bar{X} = (A+\bar{B}+\bar{C})(\bar{A}+B+\bar{C})(\bar{A}+\bar{B}+\bar{C})(\bar{A}+\bar{B}+C)(A+B+C)$

53. INVERTING BOOLEAN EQUATIONS

In Boolean algebra, as in "conventional" algebra, the same operation can be performed on both sides of an equation without destroying the equality. Thus, for example, a NOT bar can be placed over each side of a Boolean equation. Further manipulation often will lead to a simplified equation.

Example: $X = \bar{A}BC + AB\bar{C}$

$\bar{X} = \overline{\bar{A}BC + AB\bar{C}}$ (placing NOT bar over both sides of equation)

$\bar{X} = (\overline{\bar{A}BC})(\overline{AB\bar{C}})$ (De Morgan's theorem)

$\bar{X} = (\bar{\bar{A}} + \bar{B} + \bar{C})(\bar{A} + \bar{B} + \bar{\bar{C}})$ (De Morgan's theorem)

$\bar{X} = (A + \bar{B} + \bar{C})(\bar{A} + \bar{B} + C)$ (because $\bar{\bar{A}} = A$)

Comparison of the first and last equations above reveals that one has NOT bars where the other doesn't, one has OR's where the other has AND's, etc. This leads to the following rule:

Any Boolean equation can be inverted without destroying its logical meaning. The inversion is accomplished by:

1. Removing all the NOT bars from the original equation.
2. Inserting NOT bars over the variables which did not have them in the original equation.
3. Changing all AND's to OR's, and all OR's to AND's.

Assignment: Invert the following equations in accordance with the above rule. (Simplify so that no more than one NOT bar appears above any variable.) Place answers in spaces provided.

1. $X = ABC + \bar{A}\bar{B}\bar{C}$ 1. $\bar{X} =$

2. $X = A + B + C$ 2. $\bar{X} =$

3. $X = A + \bar{B}\bar{C}$ 3. $\bar{X} =$

4. $X = (A + \bar{B} + C)D$ 4. $\bar{X} =$

5. $X = \overline{(A + B)(\bar{C} + \bar{D})}$ 5. $\bar{X} =$

6. $X = \overline{AB} + C$ 6. $\bar{X} =$

7. $X = A\bar{B}C + \bar{A}B\bar{C}$ 7. $\bar{X} =$

8. $X = \overline{(A\bar{B}\bar{C})} + D$ 8. $\bar{X} =$

9. $X = (\bar{A} + B + C)(A + \bar{B} + C)$ 9. $\bar{X} =$

10. $X = (A + B)(\overline{AB})$ 10. $\bar{X} =$

Name_____ Date_____ Class_____ Grade_____

54. MINTERM AND MAXTERM EQUATIONS

A basic form of the Boolean equation is one in which each term contains all of the variables (in either inverted or noninverted form) AND'ed together, and the terms are OR'ed together. An example is shown below.

$$X = AB\bar{C} + A\bar{B}C + \bar{A}B\bar{C} + \bar{A}BC$$

This form of a Boolean equation is referred to as the *sum-of-products form,* the *standard-sum form,* or the *minterm form.* Note that the equation below is not in this form because the second term does not contain *all* variables.

$$X = A\bar{B}C + AB + \bar{A}B\bar{C}$$

It can, however, be converted to minterm form as follows:

$$X = A\bar{B}C + AB + \bar{A}B\bar{C}$$
$$X = A\bar{B}C + AB(1) + \bar{A}B\bar{C} \qquad (\text{because } AB(1) = AB)$$
$$X = A\bar{B}C + AB(C+\bar{C}) + \bar{A}B\bar{C} \qquad (\text{because } C+\bar{C} = 1)$$
$$X = A\bar{B}C + ABC + AB\bar{C} + \bar{A}B\bar{C} \qquad (\text{expanding})$$

Another basic form of the Boolean equation is one in which all variables (in either inverted or noninverted form) are OR'ed in each factor, and the factors are AND'ed. This is known as the *product-of-sums form,* the *standard-product form,* or the *maxterm form.* An example is shown below.

$$X = (\bar{A} + B + \bar{C})(\bar{A} + \bar{B} + C)(A + \bar{B} + \bar{C})$$

For every maxterm equation there is a corresponding minterm equation having the same logical meaning. Conversion from minterm to maxterm form can be accomplished as follows:

1. Write the false-output form of the minterm equation (see worksheet 52).

2. Completely invert the false-output minterm equation to obtain the true-output maxterm equation (see worksheet 53).

Conversion from maxterm to minterm form can be accomplished by reversing the above procedure; completely invert the maxterm equation, then write the true-output form of the minterm equation.

Assignment: Convert the minterm equations to maxterm form, and vice versa.

1. $X = \bar{A}B + A\bar{B}$ 1. $X =$

2. $X = ABC + \bar{A}BC + A\bar{B}C + AB\bar{C} + \bar{A}\bar{B}\bar{C}$ 2. $X =$

3. $X = (\bar{A} + \bar{B})(A + B)$ 3. $X =$

4. $X = (A + B + \bar{C})(\bar{A} + B + C)$ 4. $X =$

55. KARNAUGH MAPS

The Karnaugh map is a graphical representation of a Boolean function. Karnaugh maps for two, three, and four variables are shown below.

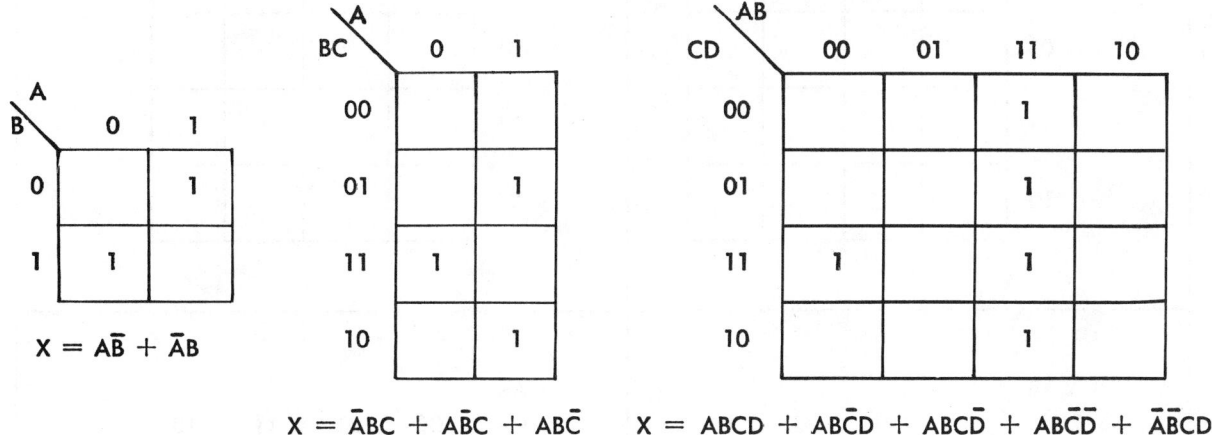

$X = A\bar{B} + \bar{A}B$

$X = \bar{A}\bar{B}C + A\bar{B}C + AB\bar{C}$

$X = ABCD + AB\bar{C}D + \bar{A}BC\bar{D} + AB\bar{C}\bar{D} + \bar{A}BCD$

Each square represents a unique combination of the input variables. As shown, a 1 is placed in each square representing a combination that will produce a *true* output (X = 1). Zeros may be placed in the remaining squares to represent *false* output (X = 0), but these zeros are often omitted to make the map easier to read.

Note that the combinations of variables listed along the edges of the map progress in the order: 00, 01, 11, and 10. A characteristic of this sequence is that only one variable changes in going from one combination to the next. Adjacent squares of the map, either horizontally or vertically, therefore differ in only one variable.

Assignment: Karnaugh maps are useful for displaying functions having up to five or six variables. Beyond that, the map becomes too complex and loses its effectiveness. A six-variable map is shown on page 114. Complete the map in accordance with the relation $X = \bar{A}\bar{B}\bar{C}\bar{D}\bar{E}\bar{F} + A\bar{B}\bar{C}\bar{D}\bar{E}\bar{F} + \bar{A}\bar{B}C\bar{D}\bar{E}\bar{F} + ABCD\bar{E}\bar{F} + A\bar{B}CD\bar{E}\bar{F} + \bar{A}BCD\bar{E}\bar{F} + A\bar{B}\bar{C}DEF + \bar{A}\bar{B}\bar{C}\bar{D}EF + \bar{A}BC\bar{D}EF + \bar{A}B\bar{C}\bar{D}EF + A\bar{B}C\bar{D}EF + AB\bar{C}\bar{D}EF$.

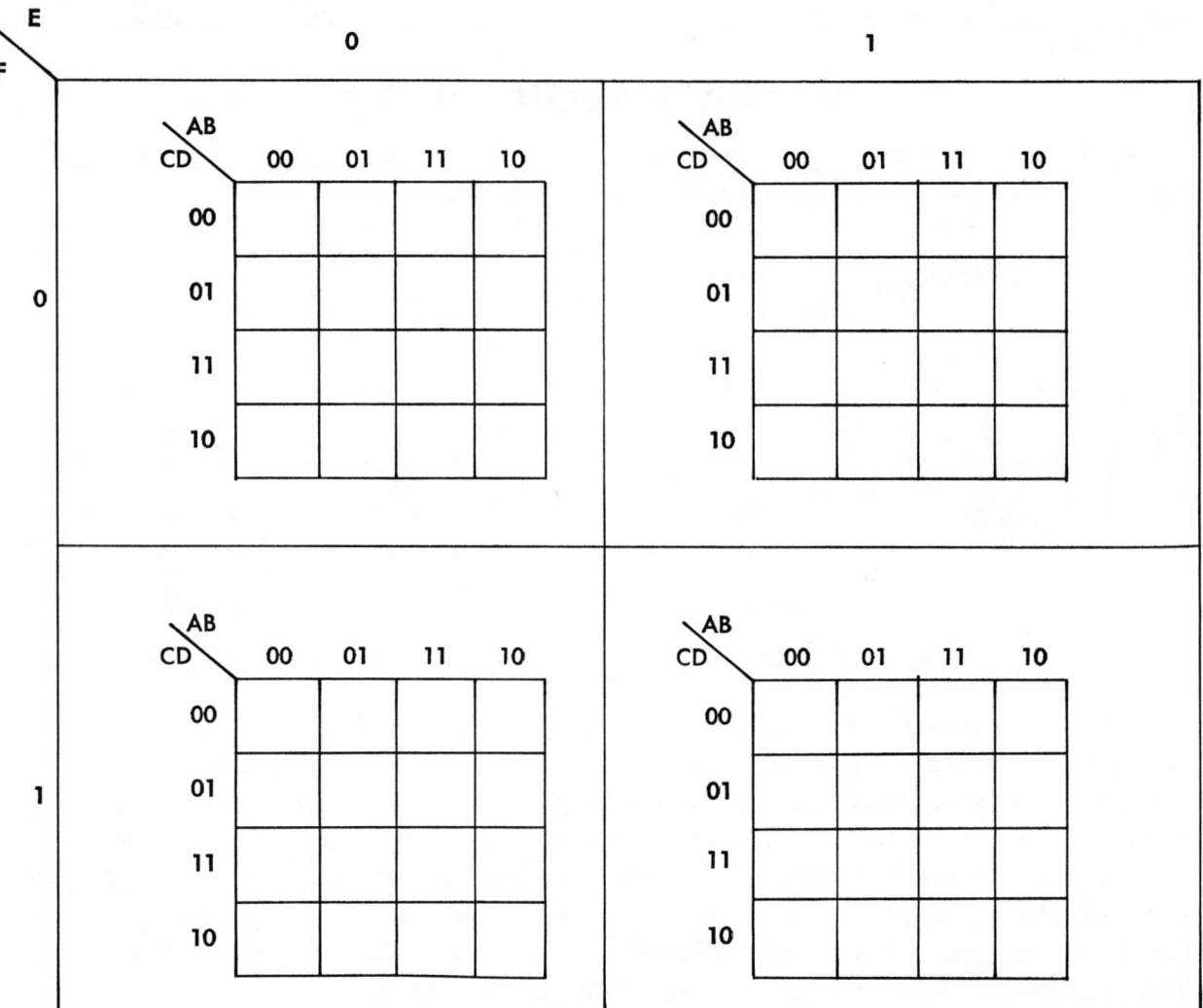

$$X = \bar{A}\bar{B}\bar{C}\bar{D}\bar{E}\bar{F} + A\bar{B}CD\bar{E}\bar{F} + A\bar{B}C\bar{D}EF + ABCDEF +$$
$$ABCD\bar{E}\bar{F} + \bar{A}BCD\bar{E}F + A\bar{B}C\bar{D}EF + \bar{A}B\bar{C}D\bar{E}F +$$
$$\bar{A}BC\bar{D}\bar{E}F + \bar{A}\bar{B}CD\bar{E}F + A\bar{B}CD\bar{E}F + AB\bar{C}D\bar{E}F$$

56. CONSTRUCTION OF KARNAUGH MAPS

For easier readability, the Karnaugh map may be drawn as shown below, with the variables themselves rather than 1's and 0's printed along the edges of the map.

When a Boolean equation is expressed in its full minterm form, there is a one-to-one correspondence between the terms of the equation and the "1" squares of the Karnaugh map. If the equation is not already in this form, it can be made so by inserting the missing terms as follows:

$$X = AC + \bar{A}BC$$
$$X = AC(B+\bar{B}) + \bar{A}BC$$
$$X = ABC + A\bar{B}C + \bar{A}BC$$

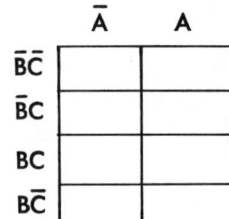

In the first equation above, the term AC implies that X is true (X = 1) whenever both A AND C are true (A = 1 AND C = 1). Whether B is true (B = 1) or false (B = 0) is of no consequence. The term AC therefore actually represents two terms: ABC + A$\bar{B}$C. The final equation above contains three terms, and there are three "1" squares in the corresponding Karnaugh map.

Assignment: Complete the following Karnaugh maps for the equations indicated.

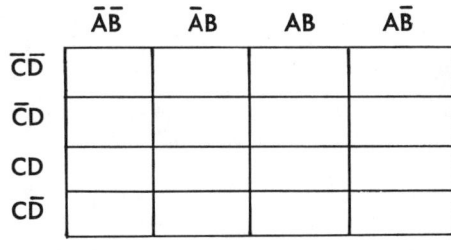

Map 1. X = $\bar{A}BC$ + $AB\bar{C}$ + $A\bar{B}C$ + ABC

Map 2. X = ABC + $\bar{A}BC$ + $\bar{B}C$

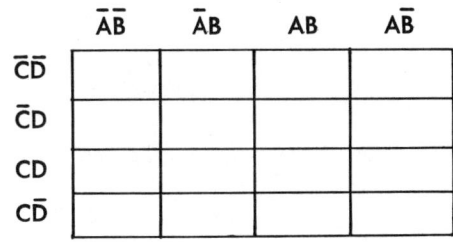

Map 3. X = ABCD + ACD + AD

Map 4. X = ABC + BD

Name________________________________Date______________Class________Grade________

57. ADJACENCIES IN KARNAUGH MAPS

In a Karnaugh map, squares containing a 1 are considered to be adjacent if they have one side in common. Furthermore, the top and bottom rows of the map are considered adjacent—as if the bottom of the map had been turned up to touch the top edge of the map. Similarly, the left and right columns are considered adjacent.

In the Karnaugh map at the left below, the adjacencies have been indicated by loops drawn around the 1's in adjacent squares. Adjacencies involving opposite edges of the map are represented as shown. The number of 1's included within a loop should be 2, 4, 8, 16, etc.; and the 1's should be in a symmetrical array. Some permissible four-loop configurations are shown at the center below. Note that the 1's are in either a two-by-two or a four-by-one arrangement. A permissible eight-loop configuration consists of a four-by-two arrangement.

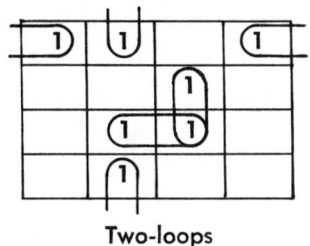

Two-loops

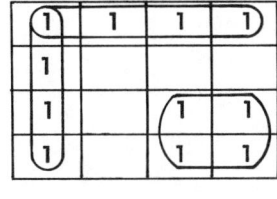

Four-loops

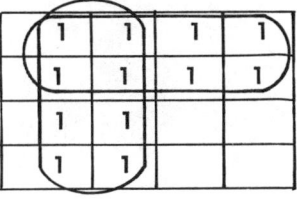

Eight-loops

Loops should be drawn in accordance with the following rules: (1) Select the largest possible configurations; e.g., an eight-loop instead of two adjacent four-loops; (2) leave as few unlooped 1's as possible; and (3) use as few loops as possible consistent with the preceding rules.

Assignment: **In accordance with the rules above, draw loops on the following maps.**

1			
1	1	1	1
		1	

Map 1

1			1
1			1
		1	
	1		1

Map 2

1	1	1	1
1			1
1	1	1	1

Map 3

		1	1
1		1	1
	1	1	

Map 4

58. READOUT OF KARNAUGH MAPS

After loops have been drawn to indicate the adjacencies in a Karnaugh map, readout of the map will yield the simplest form of the corresponding Boolean equation. In the map below, the Boolean expression represented by the two-loop is $ABCD + A\bar{B}CD$. This is then reduced to ACD as shown. A two-loop always will yield a term having one less variable than the number of variables in the map. The map below, for example, has four variables, and the two-loop yields a three-variable term (ACD). Note that readout of a loop yields a term containing only those variables which remain constant in all squares within the loop (the variable B does not appear in the readout from the two-loop because it *changes* from B to $\bar{B}$).

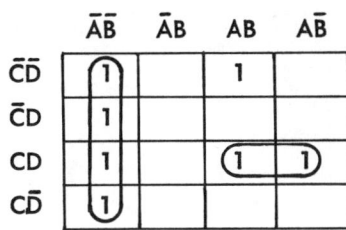

Readout of Two-Loop:

$ABCD + A\bar{B}CD = ACD (B + \bar{B})$
$= ACD$

A four-loop yields a term having two fewer variables than the number of variables in the map. The four-loop in the map above therefore yields a two-variable term:

$$\bar{A}\bar{B}\bar{C}\bar{D} + \bar{A}\bar{B}\bar{C}D + \bar{A}\bar{B}CD + \bar{A}\bar{B}C\bar{D} = \bar{A}\bar{B}(\bar{C}\bar{D} + \bar{C}D + CD + C\bar{D})$$
$$= \bar{A}\bar{B}(1)$$
$$= \bar{A}\bar{B}$$

Note again that the readout contains only those variables which remain constant throughout the loop: $\bar{A}\bar{B}$.

An eight-loop will yield a term having three fewer variables than the number of variables in the map.

Note that the above map contains an unlooped 1, and this must be accounted for during readout. The full equation corresponding to this map is: $X = \bar{A}\bar{B} + ACD + AB\bar{C}\bar{D}$.

Assignment: Loop the adjacencies in the maps, and then read out and indicate the simplest Boolean equation for each map. REMINDER: When drawing loops, remember that top and bottom rows are assumed to be adjacent, and left and right columns also are assumed to be adjacent.

	$\bar{A}\bar{B}$	$\bar{A}B$	AB	$A\bar{B}$
$\bar{C}\bar{D}$	1	1	1	
$\bar{C}D$		1	1	
CD				
$C\bar{D}$	1			

Map 1. X =

	$\bar{A}\bar{B}$	$\bar{A}B$	AB	$A\bar{B}$
$\bar{C}\bar{D}$			1	
$\bar{C}D$			1	
CD		1	1	
$C\bar{D}$	1		1	

Map 2. X =

	A̅B̅	A̅B	AB	AB̅
C̅D̅	1			1
C̅D		1		
CD		1		
CD̅	1			1

Map 3. X =

	A̅B̅	A̅B	AB	AB̅
C̅D̅	1	1	1	1
C̅D		1		
CD		1		
CD̅	1	1	1	1

Map 4. X =

59. SIMPLIFICATION OF BOOLEAN EQUATIONS BY USE OF KARNAUGH MAPS

Because adjacent squares in a Karnaugh map differ in only one variable, the variable that changes in going from one square to an adjacent square is redundant. It is for this reason that a two-loop can be read out with one variable fewer than the number of variables in the map. For a four-loop, all variables except two remain the same in all four squares. The variables which change are therefore redundant and are eliminated during readout. In an eight-loop, all variables remain the same except three which are eliminated during readout.

Assignment: For each of the following equations construct a Karnaugh map, loop the adjacencies, and then read out the simplest Boolean equation corresponding to the map.

1. $X = ABC + A\bar{B}C + \bar{A}BC$

2. $X = A\bar{B}C\bar{D} + A\bar{B}CD + \bar{A}BCD + \bar{A}\bar{B}CD$

3. $X = \bar{A}BCD + A\bar{B}CD + AB\bar{C}D + ABC\bar{D}$

4. $X = AB + \bar{A}C + BC$

5. $X = B + ABCD + \bar{A}BCD + AB\bar{C}D + ABC\bar{D}$

60. SIMPLIFICATION OF LOGIC DIAGRAMS BY USE OF KARNAUGH MAPS

Assignment: For each logic diagram below, construct a Karnaugh map, read out the simplest Boolean equation, and then draw the simplest logic diagram.

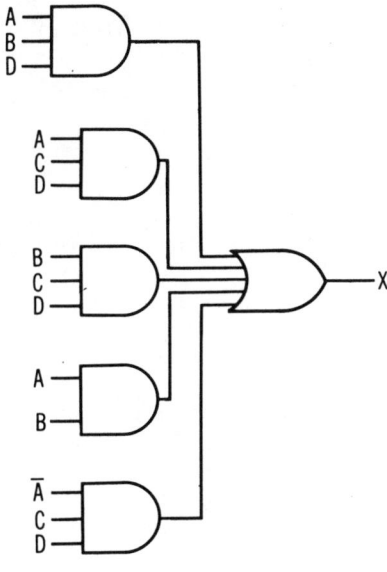

Map 1   Fig. 1

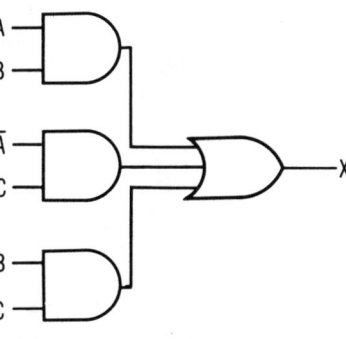

Map 2   Fig. 2

Map 3

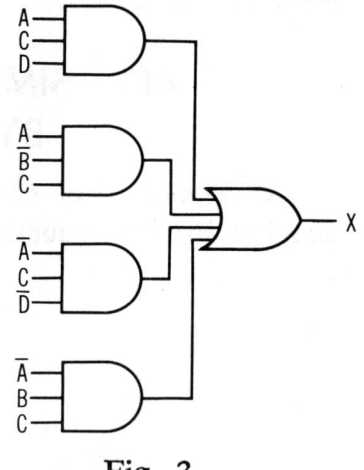

Fig. 3

Map 4

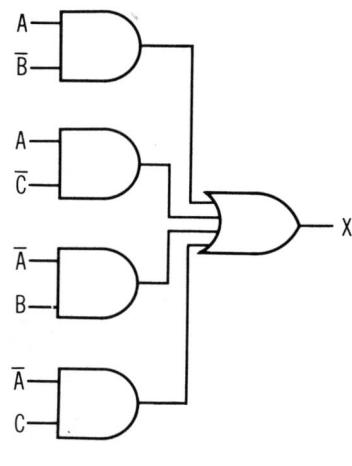

Fig. 4

Name_____Date_____Class_____Grade_____

61. INVERTING BOOLEAN EQUATIONS BY USE OF KARNAUGH MAPS

The inverse of a Boolean equation can be determined by (1) constructing a Karnaugh map of the equation, (2) constructing a second Karnaugh map with ones only in those squares containing zeros in the first map, and (3) reading out the second map. For example, supposing it is desired to find an equation for $\overline{X}$ when the equation for X is given: $X = \overline{A}\overline{B} + A\overline{C}$. The Karnaugh map for this equation is shown below. Also shown is an inverted map containing ones where the first map contains zeros. Looping and reading out the second map will yield: $\overline{X} = \overline{A}B + AC$. Actually, the solution can be reached without drawing a second Karnaugh map, by looping and reading the *zero* squares of the first map.

	$\overline{A}\overline{B}$	$\overline{A}B$	AB	$A\overline{B}$
$\overline{C}\overline{D}$	1		1	1
$\overline{C}D$	1		1	1
CD	1			
$C\overline{D}$	1			

$X = \overline{A}\overline{B} + A\overline{C}$

	$\overline{A}\overline{B}$	$\overline{A}B$	AB	$A\overline{B}$
$\overline{C}\overline{D}$		1		
$\overline{C}D$		1		
CD		1	1	1
$C\overline{D}$		1	1	1

$\overline{X} = \overline{A}B + AC$

Assignment: Using the method described above, find the equation for $\overline{X}$ in each problem below.

1. $X = A\overline{B}C\overline{D} + \overline{A}\overline{B} + A\overline{B}$

2. $X = \overline{A}B + \overline{A}BC + \overline{A}B\overline{C}D$

3. $X = A + ABCD + A\bar{B}D$

4. $X = \bar{A}\bar{B}\bar{C}\bar{D} + A\bar{B}CD + A\bar{B}C\bar{D} + \bar{A}\bar{B}\bar{C}D$

62. THE VEITCH DIAGRAM

Veitch diagrams, like Karnaugh maps, are graphical representations of a logical relationship, and are useful for minimization of Boolean equations. The Veitch diagram differs from the Karnaugh map primarily in the order in which the variables are assigned to the squares. A Veitch diagram for four variables is shown below, and is constructed for the equation $X = \bar{A}BCD + \bar{A}BC\bar{D} + \bar{A}\bar{B}CD + \bar{A}\bar{B}C\bar{D}$. Readout of this diagram yields: $X = \bar{A}C$. Readout of a Veitch diagram is similar to readout of a Karnaugh map; the adjacencies, sometimes called *couples*, are looped, and the simplest equation is determined from the loops.

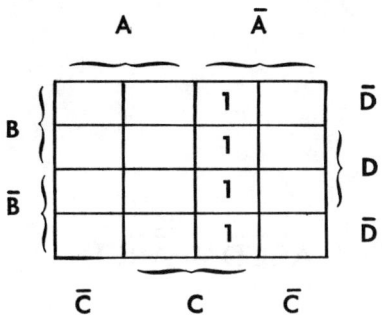

Veitch diagram for $X = \bar{A}C$

Assignment: For each equation below, construct a Veitch diagram and read out the minimum equation.

1. $X = ABC + ABD + \bar{A}\bar{B}CD + ABC\bar{D}$

2. $X = A\bar{B}\bar{C}D + ABCD + A\bar{B}CD + AB\bar{C}D + \bar{A}BC\bar{D}$

3. $X = A + B + ACD + ABD + AB\bar{C}\bar{D}$

4. $X = \bar{B}CD + A\bar{B}\bar{C}D + A\bar{B}\bar{C}\bar{D} + \bar{A}BCD$

63. ADDITIONAL NOTES ON LOGIC CIRCUITS

EXCLUSIVE OR: The conventional OR gate, shown below, produces a binary 1 output if there is a binary 1 input at *either or both* of its input terminals. Because of this "either or both" characteristic, this gate is sometimes referred to as an *inclusive* OR to distinguish it from the *exclusive* OR. The exclusive OR produces a binary 1 output if "either but not both" inputs are binary 1. The underlying logic, along with the single-symbol representation for the exclusive OR, are shown below.

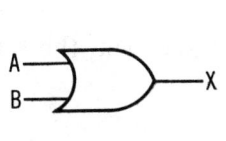

Inclusive OR

A	B	X
0	0	0
0	1	1
1	0	1
1	1	1

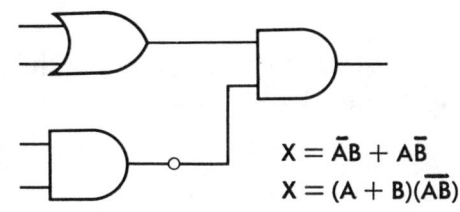

$X = \bar{A}B + A\bar{B}$
$X = (A + B)(\overline{AB})$

Underlying Logic of Exclusive OR

Exclusive OR

A	B	X
0	0	0
0	1	1
1	0	1
1	1	0

NAND GATE: The NAND gate and its truth table are shown below. As the symbol suggests, the NAND behaves as an AND gate followed by an inverter. This gate will produce a binary 0 output *only* when both inputs are binary 1. Any other combination of inputs will produce a binary 1 output.

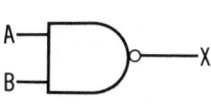

NAND Gate

A	B	X
0	0	1
0	1	1
1	0	1
1	1	0

NOR GATE: The NOR gate and its truth table are shown below. As the symbol suggests, the NOR gate behaves as an OR gate followed by an inverter. This gate produces a binary 1 output *only* when both inputs are binary 0, i.e., output is 1 when *neither A nor B are* 1.

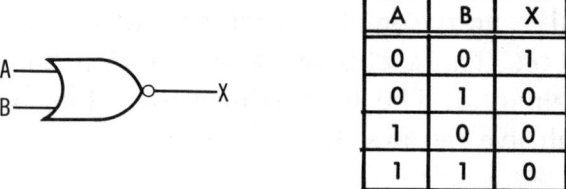

A	B	X
0	0	1
0	1	0
1	0	0
1	1	0

Assignment: Complete the following Venn diagrams for the EXCLUSIVE OR, the NAND gate and the NOR gate.

Exclusive OR

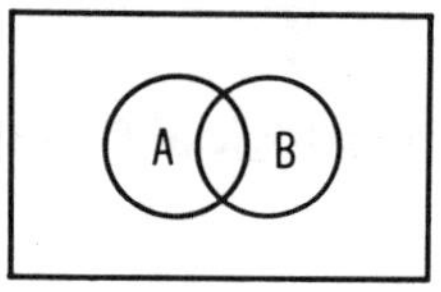

Fig. 1

NAND

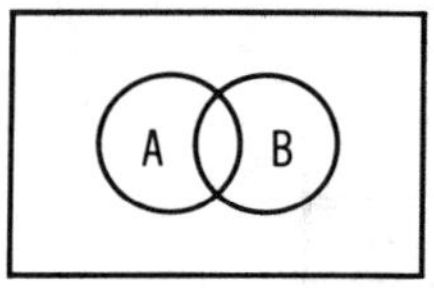

Fig. 2

NOR

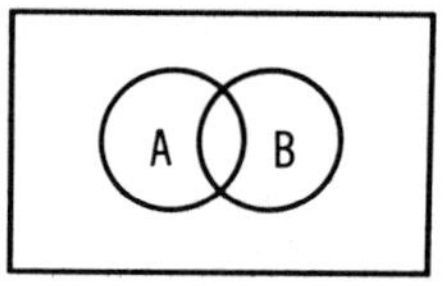

Fig. 3

64. ADDITIONAL NOTES ON LOGIC SYMBOLS

Attempts have been made to standardize the symbols used to represent logic circuits. To date, by far the most widely accepted symbols are those specified in military standard MIL STD 806 (the symbols used in this workbook).

Another standard, generated by the Institute of Electrical and Electronic Engineers (IEEE), is less widely accepted. This standard specifies two types of symbols: rectangular blocks and distinctively shaped symbols as in MIL STD 806. An inscription inside each block identifies its function, i.e., *and, or,* etc. As shown below, the ampersand (&) identifies the AND gate, and the symbol for "equal to or greater than one" identifies the OR gate (because output is active when one or more inputs are active). The *exclusive* OR is identified by the symbol =1, meaning that the output is active when exactly one input is active.

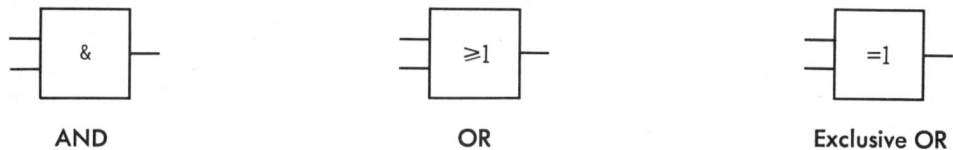

      AND   OR   Exclusive OR

The IEEE standard is generally favored by those who prefer to think of inputs/outputs as either "high" or "low" (more positive or less positive) rather than in terms of binary 1's and 0's. The symbols reflect this preference. Rather than using the small circle to indicate *negation* or *inversion,* as in MIL STD 806, the IEEE symbols employ a triangle to mean "active when low." Triangles may appear on input or output lines of a logic gate. The absence of a triangle on a given line means "active when high." Therefore, in the symbol below, the output will be low only when both inputs A and B are high.

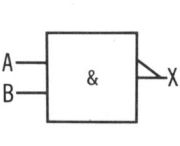

A	B	X
L	L	H
L	H	H
H	L	H
H	H	L

Assignment: Complete the truth table for the gate below.

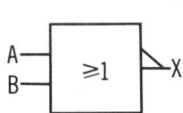

A	B	X
L	L	
L	H	
H	L	
H	H	

133

Worksheet 1
1. $8^4 = 4096$
2. $5^{-2} = 1/25$
3. $26^2 = 676$
4. octal 15
5. trinary 222

Worksheet 2
1. baa
2. baaa
3. 100_8
4. 10_2
5. 100000_2

Worksheet 3
1. 4103_{10}
2. 64_{10}
3. 668_{10}
4. 511_{10}
5. 2137_{10}
6. 22_{10}
7. 52_{10}
8. 361_{10}
9. 50_{10}
10. 37.625_{10}

Worksheet 4
1. 33_8
2. 1173_8
3. 5102_8
4. 1757_8
5. 1611_8
6. 1000_8
7. 710_8
8. 21270_8
9. 316_8
10. 7777_8

Worksheet 5
1. $.5000_8$
2. $.0753_8$
3. $.600_8$
4. $.031_8$
5. $.200_8$
6. $.00040_8$
7. $.7773_8$
8. $.40_8$
9. $.63_8$
10. 326.243_8

Worksheet 6
1. 5_{10}
2. $.125_{10}$
3. 15_{10}
4. 158_{10}
5. 21_{10}
6. 45_{10}
7. 254_{10}
8. 1100100_2
9. 1110000_2
10. 1000000_2

Worksheet 7
1. $.625_{10}$
2. $.125_{10}$
3. $.7500_{10}$
4. $.875_{10}$
5. $.5325_{10}$
6. $.3125_{10}$
7. $.1875_{10}$
8. $.6679_{10}$
9. $.984_{10}$
10. $.3808_{10}$

Worksheet 8
1. 11001_2
2. 110110_2
3. 11011000_2
4. 100010_2
5. 10000000_2
6. 100000110_2
7. 100010010_2
8. 100111001_2
9. 10000000000_2
10. 100101100000_2

Worksheet 9
1. $.100_2$
2. $.01010100_2$
3. $.0000001_2$
4. $.1111111_2$
5. $.01000_2$
6. $.00101010_2$
7. $.0000000000000000001_2$
8. $.11000000_2$
9. $.000111000111_2$
10. 11001.010000_2

Worksheet 10
1. 33_{10}
2. 27_{10}
3. 63_{10}
4. 682_{10}
5. 8_{10}
6. 365_{10}
7. 19_{10}
8. 240_{10}
9. 36_{10}
10. 924_{10}

Worksheet 11
1. 1110_2
2. 35_8
3. 100_8
4. 111111111_2
5. 6222_8
6. 33_8
7. 1011100.110000101_2
8. 100100100_2
9. 6.71_8
10. 10000000000000_2

Worksheet 12
1. 0001 0010 0111
2. 1001 1001 . 0111 0110
3. 0001 0010 0000 0000
4. 0110 . 0010 1000
5. 0001 0011 0101 0111 – 0110 0100 0010 1000

Worksheet 13
1. 0101 0011 1100
2. 1010 0111 1001 0110
3. 1100 0011 1100 1000 0111
4. 0100 0011 0011 0100
5. 1011 0100

Worksheet 14
1. 00011 00101 01100
2. 01001 00110 01001 01100
3. 00011 00110 00110
4. 01100 01010 01001
5. 10001 01001 00101 00011 – 00110

Worksheet 18
1. 100
2. 110
3. 10
4. 100
5. 1111
6. 1111
7. 10001
8. 11
9. 100
10. 11.011

Worksheet 19
1. 85726_{10}
2. 49997_{10}
3. 010101_2
4. 501_8
5. 1020_3
6. 736_{10}
7. 3700_{10}
8. 544_8
9. 010_3
10. 00011_2

Worksheet 20
1. 62
2. 407
3. 364
4. 104
5. 274
6. 20
7. 1
8. 145
9. 60
10. 156

Worksheet 21
1. 10010
2. 101
3. 100000
4. 10101
5. −1
6. 100
7. 100
8. 1101110
9. 11011
10. −10

Worksheet 22
1. 11110
2. 110001
3. 100100010
4. 100101001
5. 101000.1001
6. 101
7. 101
8. 101
9. 110
10. 100001.0101

Worksheet 23
1. 367_8
2. 43707_8
3. 10000_8
4. 1776_8
5. 160_8

Octal Addition Table

0	0	1	2	3	4	5	6	7
1	1	2	3	4	5	6	7	10
2	2	3	4	5	6	7	10	11
3	3	4	5	6	7	10	11	12
4	4	5	6	7	10	11	12	13
5	5	6	7	10	11	12	13	14
6	6	7	10	11	12	13	14	15
7	7	10	11	12	13	14	15	16
	0	1	2	3	4	5	6	7

Worksheet 24
1. 651_8
2. 6771_8
3. 2555374_8
4. 203_8
5. 1.463_8

Octal Multiplication Table

0	0	0	0	0	0	0	0	0
1	0	1	2	3	4	5	6	7
2	0	2	4	6	10	12	14	16
3	0	3	6	11	14	17	22	25
4	0	4	10	14	20	24	30	34
5	0	5	12	17	24	31	36	43
6	0	6	14	22	30	36	44	52
7	0	7	16	25	34	43	52	61
	0	1	2	3	4	5	6	7

Worksheet 25
1. 011, 012, 013, 014, 015, 016, 017, 018, 019, 01A, 01B, 01C, 01D, 01E, 01F, 020, 021, 022, 023, 024, 025, 026, 027, 028, 029, 02A, 02B, 02C, 02D, 02E, 02F
2. 1000_{16}
3. 256_{10}
4. 168_{16}
5. 64_{16}
6. 640_{16}
7. 2748_{10}
8. 110000011100102
9. 0011 0000 0111 0010
10. 111110011000110001112

Worksheet 26

EQUATION	FIGURE
I	2
II	3
III	1
IV	4

Worksheet 27
2. $D = ABC$
6. $H = D + EF$
7. $G = EF$
8. $H = D + G$
10. $H = EF + ABC$

Worksheet 28
3. $J = G(HI)$
5. $G = A \lor B \lor C \lor D$
7. $H = E + F$
8.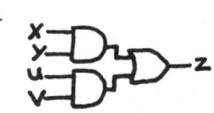

$L = MN + P$

9.

$Z = (x \cdot y) + (u \cdot v)$

10.

$E = (A+B)(C+D)$

Worksheet 29

EQUATION	FIGURE
I	3
II	1
III	4
IV	2

Worksheet 30

EQUATION	FIGURE
I	4
II	3
III	1
IV	2
V	5

Worksheet 31

1. 1
2. A
3. 0
4. 1
5. C
6. A
7. 1
8. 1
9. 1
10. 0

Worksheet 32

1. 1
2. 1
3. 1
4. $A + BC$
5. $A + B + C$
6. E
7. B
8. DE
9. BC
10. $\bar{A}B + C$

Worksheet 33

A	B	C	ABC	A+B+C	$\overline{AB}\bar{C}$	$\bar{A}B + C$
0	0	0	0	0	1	1
0	0	1	0	1	0	1
0	1	0	0	1	0	0
0	1	1	0	1	0	1
1	0	0	0	1	0	0
1	0	1	0	1	0	1
1	1	0	0	1	0	0
1	1	1	1	1	0	1

Worksheet 34

Truth Table 1

A	B	C	AC	$\bar{B}$	$AC + \bar{B}$
0	0	0	0	1	1
0	0	1	0	1	1
0	1	0	0	0	0
0	1	1	0	0	0
1	0	0	0	1	1
1	0	1	1	1	1
1	1	0	0	0	0
1	1	1	1	0	1

Truth Table 2

A	B	C	$\bar{A}$	$\bar{A}B$	$\bar{A}B + C$
0	0	0	1	0	0
0	0	1	1	0	1
0	1	0	1	1	1
0	1	1	1	1	1
1	0	0	0	0	0
1	0	1	0	0	1
1	1	0	0	0	0
1	1	1	0	0	1

Truth Table 3

A	B	C	$\bar{B}$	$A + \bar{B} + C$
0	0	0	1	1
0	0	1	1	1
0	1	0	0	0
0	1	1	0	1
1	0	0	1	1
1	0	1	1	1
1	1	0	0	1
1	1	1	0	1

Truth Table 4

A	B	C	$\bar{B}$	$\bar{C}$	$A + \bar{B}\bar{C}$
0	0	0	1	1	1
0	0	1	1	0	0
0	1	0	0	1	0
0	1	1	0	0	0
1	0	0	1	1	1
1	0	1	1	0	1
1	1	0	0	1	1
1	1	1	0	0	1

Worksheet 35

Truth Table 1

A	B	A+B	A(A+B)
0	0	0	0
0	1	1	0
1	0	1	1
1	1	1	1

$A(A+B) = A$

Truth Table 2

A	B	$\bar{A}$	$\bar{A}B$	$A + \bar{A}B$	A+B
0	0	1	0	0	0
0	1	1	1	1	1
1	0	0	0	1	1
1	1	0	0	1	1

$A + \bar{A}B = A + B$

Truth Table 3

A	B	C	B+C	A(B+C)	AB	AC	AB+AC
0	0	0	0	0	0	0	0
0	0	1	1	0	0	0	0
0	1	0	1	0	0	0	0
0	1	1	1	0	0	0	0
1	0	0	0	0	0	0	0
1	0	1	1	1	0	1	1
1	1	0	1	1	1	0	1
1	1	1	1	1	1	1	1

$A(B+C) = AB + AC$

Truth Table 4

A	B	C	BC	A+BC	A+B	A+C	(A+B)(A+C)
0	0	0	0	0	0	0	0
0	0	1	0	0	0	1	0
0	1	0	0	0	1	0	0
0	1	1	1	1	1	1	1
1	0	0	0	1	1	1	1
1	0	1	0	1	1	1	1
1	1	0	0	1	1	1	1
1	1	1	1	1	1	1	1

$A + BC = (A+B)(A+C)$

Worksheet 36

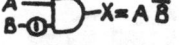

Fig. 1.

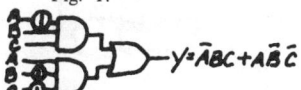

Fig. 2.

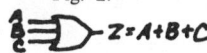

Fig. 3.

Fig. 4.

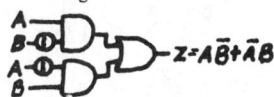

Fig. 5.

Worksheet 37

Fig. 1. $X = AB + BC$ (or $X = B(A+C)$)
Fig. 2. $X = \bar{A}\bar{B}C$
Fig. 3. $X = \bar{A} + B + C$
Fig. 4. $X = (A+B)(A+C)$
Fig. 5. $X = (A+B)(\overline{AB})$ (or ($X = A\bar{B} + \bar{A}B$))

Worksheet 38

Table 1

A	B	X
0	0	1
0	1	0
1	0	1
1	1	1

Table 2

A	B	X
0	0	0
0	1	0
1	0	1
1	1	0

Table 3

A	B	C	X
0	0	0	0
0	0	1	0
0	1	0	0
0	1	1	1
1	0	0	1
1	0	1	1
1	1	0	1
1	1	1	1

Table 4

A	B	X
0	0	0
0	1	1
1	0	1
1	1	0

Table 5

A	B	X
0	0	1
0	1	1
1	0	1
1	1	0

Worksheet 39

Fig. 1.

Fig. 2.

Fig. 3.

Fig. 4.

Worksheet 40

Table 1

A	B	X
0	0	1
0	1	0
1	0	0
1	1	0

Table 2

A	B	X
0	0	0
0	1	1
1	0	1
1	1	1

Table 3

A	B	C	X
0	0	0	0
0	0	1	0
0	1	0	0
0	1	1	1
1	0	0	0
1	0	1	1
1	1	0	1
1	1	1	1

Table 4

A	B	X
0	0	1
0	1	0
1	0	0
1	1	1

Table 5

A	B	X
0	0	0
0	1	1
1	0	1
1	1	0

Worksheet 41

Eq. 1. $X = A + B$
Eq. 2. $X = A\bar{B}$
Eq. 3. $X = \bar{A}B + A\bar{B}$
Eq. 4. $X = \overline{AB} + \bar{A}B + A\bar{B}$ (or $X = \overline{AB}$)
Eq. 5. $X = \bar{A}\bar{B}$ (or $X = \overline{A+B}$)

Worksheet 42

Fig. 1. $X = ABC$

Fig. 2. $X = A\bar{B}$

Fig. 3. $X = \bar{A}\bar{B}$

Fig. 4. $X = \bar{A} + BC$

Worksheet 43

1. $X = A\bar{B} + \bar{A}B$
2. $X = \bar{A}B$
3. $X = AB + \bar{A}\bar{B}$
4. $X = C(A+B) = AC + BC$
5. $X = \bar{A}\bar{B} + A\bar{B} + \bar{A}B = \bar{A}\bar{B}$

Worksheet 44

Fig. 1. $A\bar{B}$ Fig. 2. $\bar{A} + B$

Fig. 3. $C(A\bar{B} + \bar{A}B)$

Fig. 3. Alternate. $A\bar{B}C + \bar{A}BC$

Fig. 4. $A\bar{B} + \bar{A}B$

Fig. 5. $\overline{AB}$

Worksheet 45

Fig. 1.

Fig. 2.

Fig. 3.

Fig. 4.

Fig. 5.

Worksheet 46

1. $A\bar{B} + AB + \bar{A}B = A\bar{B} + AB + \bar{A}B + AB$
$= A(\bar{B} + B) + B(\bar{A} + A)$
$= A(1) + B(1)$
$= A + B$

2. $A\bar{B} + AB + \bar{A}B$ $A + B$

3.

A	B	X
0	0	0
0	1	1
1	0	1
1	1	1

A	B	X
0	0	0
0	1	1
1	0	1
1	1	1

Worksheet 47

1. A
2. B
3. 1
4. $C(A+B)$
5. ABC

Worksheet 48

Fig. 1. $X = AB$

Fig. 2. $X = C(A+B)$

Fig. 3. $X = A\bar{D}$

Fig. 4. $X = A+B$

2. $\bar{X} = \bar{A}\bar{B}\bar{C}$
3. $\bar{X} = \bar{A}(B+C)$
4. $\bar{X} = \bar{A}B\bar{C} + \bar{D}$
5. $\bar{X} = (A+B)(\bar{C}+\bar{D})$
6. $\bar{X} = \bar{A} + \bar{B} + C$
7. $\bar{X} = (\bar{A}+B+\bar{C})(A+\bar{B}+C)$
8. $\bar{X} = A\bar{B}\bar{C} + D$
9. $\bar{X} = AB\bar{C} + \bar{A}B\bar{C}$
10. $\bar{X} = \bar{A}\bar{B} + AB$

Worksheet 49

A	B	A+B	$\overline{A+B}$	$\bar{A}$	$\bar{B}$	$\bar{A}+\bar{B}$
0	0	0	1	1	1	1
0	1	1	0	1	0	1
1	0	1	0	0	1	1
1	1	1	0	0	0	0

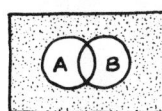
$\overline{A+B}$

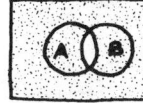
$\bar{A}+\bar{B}$

Worksheet 50

A	B	C	A+B+C	$\overline{A+B+C}$	$\bar{A}$	$\bar{B}$	$\bar{C}$	$\overline{ABC}$
0	0	0	0	1	1	1	1	1
0	0	1	1	0	1	1	0	0
0	1	0	1	0	1	0	1	0
0	1	1	1	0	1	0	0	0
1	0	0	1	0	0	1	1	0
1	0	1	1	0	0	1	0	0
1	1	0	1	0	0	0	1	0
1	1	1	1	0	0	0	0	0

Worksheet 51

1. $\bar{A} + \bar{B} + \bar{C} + \bar{D}$
2. $\bar{A} + \bar{B} + \bar{C} + \bar{D}$
3. $\bar{A}\bar{C}\bar{B}$
4. $\bar{A} + \bar{B} + \bar{C} + \bar{D}\bar{E}$
5. $(\bar{A}+\bar{B}+\bar{C})D$
6. $\overline{ABCD}$
7. $\overline{(A+B+C)(D+E)}$
8. $\overline{AC+B}$
9. $\overline{(A+B+C)DE}$
10. $(\overline{A+B+C})D$

Worksheet 52

1. $\bar{X} = AB + \bar{A}\bar{B}$
2. $\bar{X} = \bar{A}\bar{B}$
3. $\bar{X} = \bar{A}BC + A\bar{B}C + AB\bar{C} + ABC$
4. $\bar{X} = \bar{A}BC + \bar{A}B\bar{C} + A\bar{B}\bar{C}$

Worksheet 53

1. $\bar{X} = (\bar{A}+\bar{B}+\bar{C})(A+B+C)$

Worksheet 54

1. $X = (A+B)(\bar{A}+\bar{B})$
2. $X = (A+B+\bar{C})(A+\bar{B}+C)(\bar{A}+B+C)$
3. $X = A\bar{B} + \bar{A}B$
4. $X = \bar{A}BC + A\bar{B}C + AB\bar{C} + \bar{A}\bar{B}C + \bar{A}B\bar{C} + ABC$

Worksheet 55

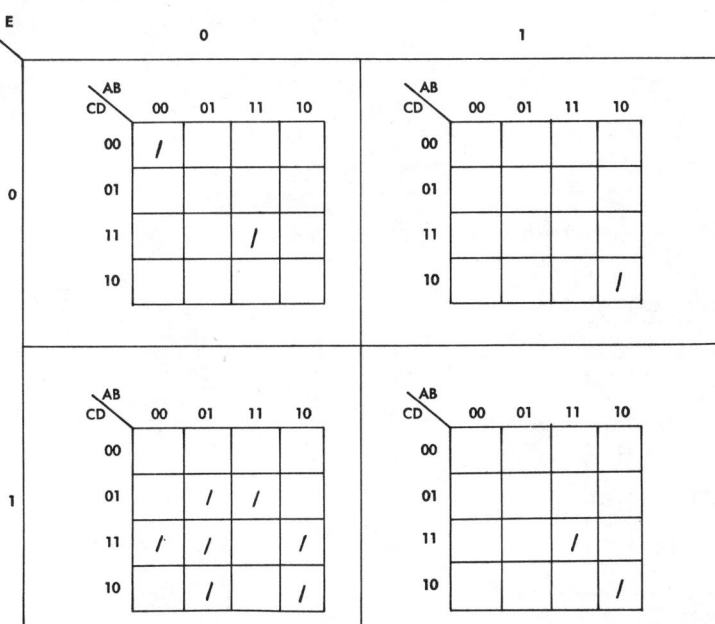

Worksheet 56

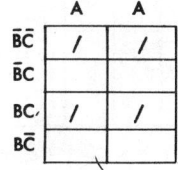

Map 1. $X = \bar{A}BC + AB\bar{C} + A\bar{B}C + ABC$

Map 2. $X = ABC + \bar{A}BC + \bar{B}C$

Map 3. $X = ABCD + ACD + AD$

Map 4. $X = ABC + BD$

Worksheet 57

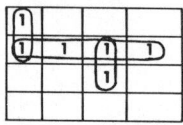
Map 1.

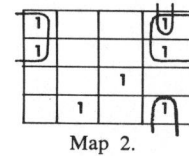
Map 2.

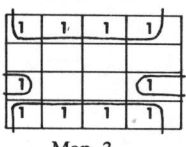
Map 3.

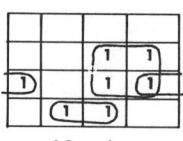
Map 4.

Worksheet 58

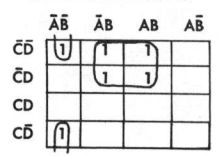
Map 1. $X = B\bar{C} + \bar{A}\bar{B}\bar{D}$

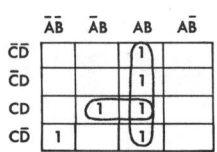
Map 2. $X = AB + BCD + \bar{A}\bar{B}C\bar{D}$

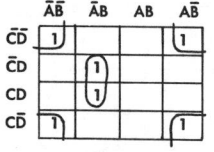
Map 3. $X = \bar{B}\bar{D} + \bar{A}BD$

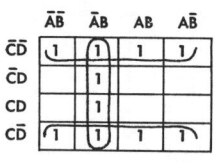
Map 4. $X = \bar{D} + \bar{A}B$

Worksheet 59

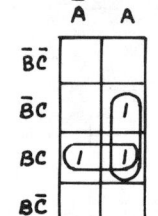 or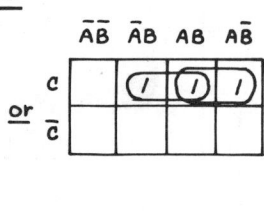

Map 1. $X = AC + BC$

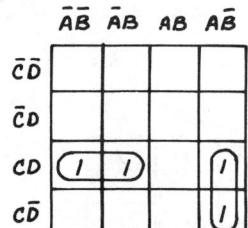
Map 2. $X = \bar{A}CD + ABC$

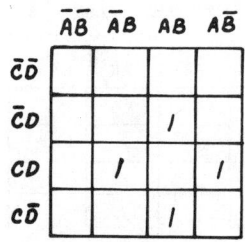
Map 3. $X = \bar{A}BCD + A\bar{B}CD + AB\bar{C}D + ABCD$

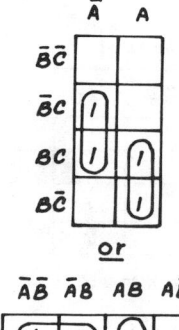

or

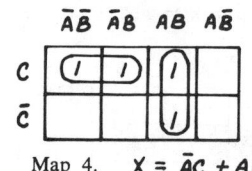
Map 4. $X = \bar{A}C + AB$

Worksheet 60

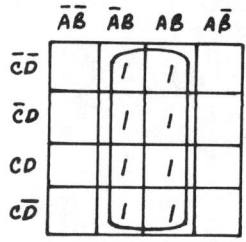
Map 5. $X = B$

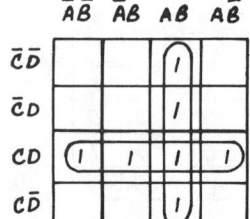
Map 1. $X = AB + CD$

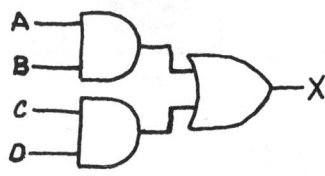
Fig. 1. $X = AB + CD$

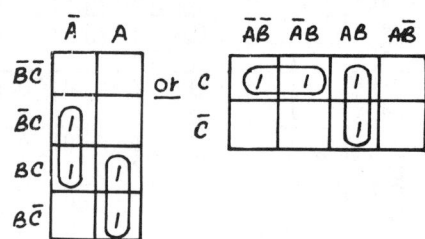
Map 2. $X = \bar{A}C + AB$

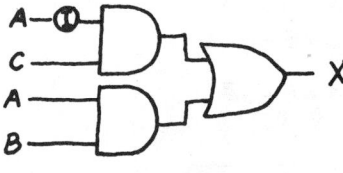
Fig. 2. $X = \bar{A}C + AB$

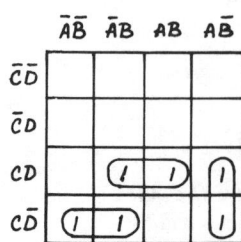
Map 3. $X = A\bar{B}C + BCD + \bar{A}C\bar{D}$

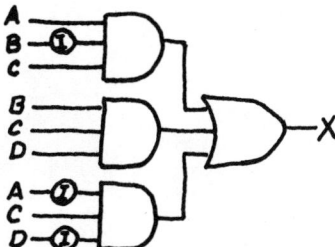

Fig. 3. $X = A\bar{B}C + BCD + \bar{A}C\bar{D}$

Alternate Solution for Map 3:

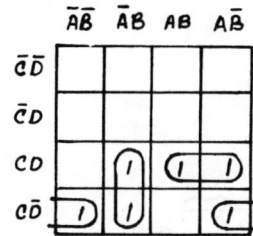

Map 3. $X = ACD + \bar{A}BC + \bar{B}C\bar{D}$

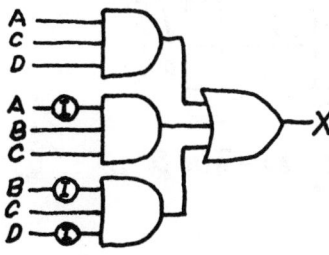

Fig. 3. $X = ACD + \bar{A}BC + \bar{B}C\bar{D}$

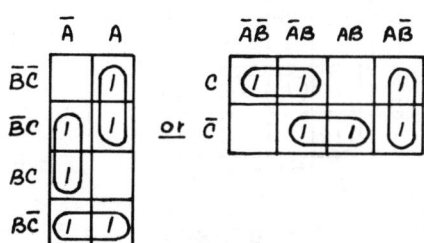

Map 4. $X = A\bar{B} + \bar{A}C + B\bar{C}$

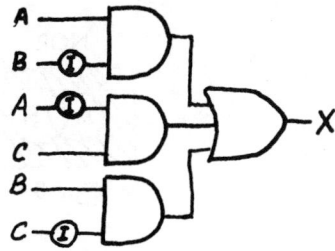

Fig. 4. $X = A\bar{B} + \bar{A}C + B\bar{C}$

Alternate Solution for Map 4:

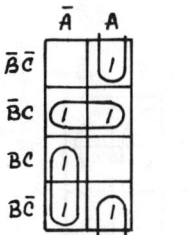

Map 4. $X = A\bar{C} + \bar{B}C + \bar{A}B$

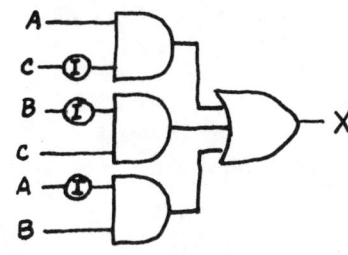

Fig. 4. $X = A\bar{C} + \bar{B}C + \bar{A}B$

Worksheet 61

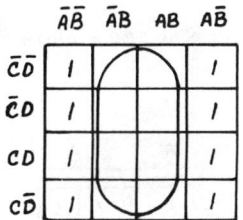

1. $X = AB\bar{C}\bar{D} + \bar{A}\bar{B} + A\bar{B}$
   $\bar{X} = B$

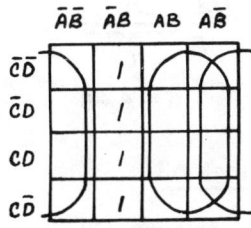

2. $X = \bar{A}B + \bar{A}BC + \bar{A}B\bar{C}D$
   $\bar{X} = A + \bar{B}$

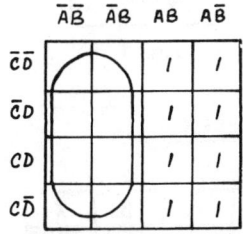

3. $X = A + ABCD + A\bar{B}D$
   $\bar{X} = \bar{A}$

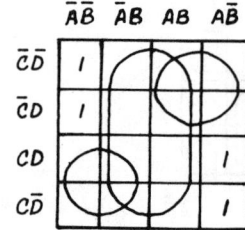

4. $X = \bar{A}\bar{B}\bar{C}\bar{D} + A\bar{B}CD + \bar{A}BC\bar{D} + \bar{A}\bar{B}\bar{C}D$
   $\bar{X} = B + \bar{A}C + A\bar{C}$

Worksheet 62

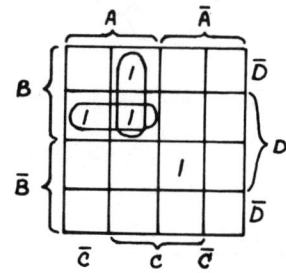

1. $X = ABC + ABD + \bar{A}BCD$

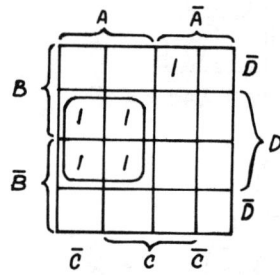

2. $X = AD + \bar{A}BC\bar{D}$

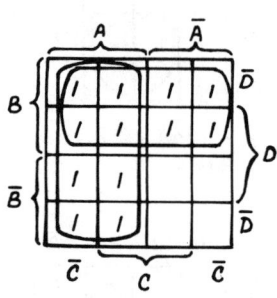

3. $X = A + B$

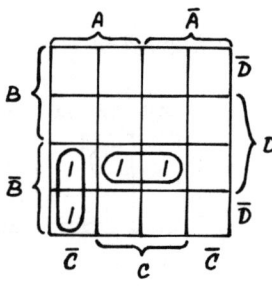

4. $X = A\bar{B}\bar{C} + \bar{B}CD$

Worksheet 63

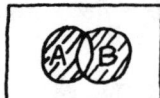

Fig. 1
Exclusive OR

Fig. 2
NAND

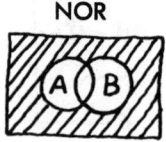

Fig. 3
NOR

Worksheet 64

A	B	X
L	L	H
L	H	L
H	L	L
H	H	L